Adaptations of Desert Organisms

Edited by J.L. Cloudsley-Thompson

Springer
Berlin
Heidelberg
New York
Barcelona
Budapest
Hong Kong
London
Milan
Paris
Santa Clara
Singapore
Tokyo

Volumes already published

Ecophysiology of the Camelidae and Desert Ruminants
By R.T. Wilson (1989)

Ecophysiology of Desert Arthropods and Reptiles
By J.L. Cloudsley-Thompson (1991)

Plant Nutrients in Desert Environments
By A. Day and K. Ludeke (1993)

Seed Germination in Desert Plants
By Y. Gutterman (1993)

Behavioural Adaptations of Desert Animals
By G. Costa (1995)

Invertebrates in Hot and Cold Arid Environments
By L. Sømme (1995)

Energetics of Desert Invertebrates
By H. Heatwole (1996)

Ecophysiology of Desert Birds
By G.L. Maclean (1996)

Plants of Desert Dunes
By A. Danin (1996)

Biotic Interactions in Arid Lands
By J.L. Cloudsley-Thompson (1996)

Structure-Function Relations of Warm Desert Plants
By A.C. Gibson (1996)

Physiological Ecology of North American Desert Plants
By S.D. Smith, R.K. Manson, and J.E. Anderson (1997)

Ecophysiology of Small Desert Mammals
By A.A. Degen (1997)

Homeostasis in Desert Reptiles
By S.D. Bradshaw (1997)

Ecophysiology of Amphibians Inhabiting Xeric Environments
By M.R. Warburg (1997)

Avian Desert Predators
By W.E. Cook (1997)

Carnivore Ecology in Arid Lands
By J. du P. Bothma (1998)

In preparation

Ecophysiology of Economic Plants in Arid and Semi-Arid Lands
By G.E. Wickens (1998)

Arthropods of Mediterranean-Type Ecosystems
By G.P. Stamou (1998)

Jacobus du P. Bothma

Carnivore Ecology in Arid Lands

With 18 Figures

Springer

Prof. Dr. Jacobus du P. Bothma
Centre for Wildlife Management
University of Pretoria
0002 Pretoria
Republic of South Africa

Front cover illustration:
Panthera pardus (leopard). Photograph taken by J. du P. Bothma.

ISSN 1430-9432
ISBN 3-540-59265-2 Springer-Verlag Berlin Heidelberg New York

Library of Congress Cataloging-in-Publication Data
Bothma, J. du P.
Carnivore ecology in arid lands / Jacobus du P. Bothma.
p. cm. -- (Adaptations of desert organisms)
Includes bibliographical referencess (p.) and index.
ISBN 3-540-59265-2 (hardcover : alk. paper)
1.Carnivore--Ecology. 2. Arid regions animals--Ecology.
3. Carnivora--Adaptation. 4. Arid regions animals--Addaptation.
I. Titel. II. Series.
QL737.C2B68 1998
599.71754--dc21

Cover design: Design & Production GmbH, Heidelberg
Camera ready by: Emanuel Rachl

SPIN 10480341 31/3137 - 5 4 3 2 1 0 - Printed on acid-free paper

Acknowledgements

Illustrations from the following sources have been used with kind permission and are gratefully acknowledged: Figure 2.4 from Fanshawe et al. (1991); Fig. 3.3 from Stander (1992a); Fig. 3.4 from Cooper (1991) and Fig. 4.2 from Tilson et al. (1980); Fig. 4.3 from Richardson (1987b); Fig. 4.4 from Richardson (1987a) and Fig. 5.2 from Rasa (1989a). Other Figures are photographs by the author.

Sincere appreciation is also expressed to Springer-Verlag GmbH & Co. KG for involving me in this Series on the Adaptations of Desert Organisms, and special thanks to Dr. Andrea Schlitzberger for her kind assistance. Prof. J.L. Cloudsley-Thompson has been an inspiration as a desert biologist over many years. He acted as a kind advisor during the preparation of this manuscript. His comments were invaluable. My wife Babsie typed and prepared many drafts of the manuscript. Prof. F.C. Eloff first introduced me to desert ecology and commented on various drafts. The University of Pretoria allowed me time to prepare the manuscript and helped in the preparation of the illustrations used in the text. All these people and institutions are thanked warmly and most sincerely.

Pretoria, December 1997 *J. du P. Bothma*

Contents

CHAPTER 1

Introduction

The members of the order Carnivora were thought originally to be flesh-eaters only, hence their name which is derived from the Latin words *caro* (flesh) and *voro* (to eat). Now we know that most carnivores are in reality omnivores (Smithers 1983). Carnivores occur widely in the world Some species may occupy diverse habitats, while others are specialized for life in more restricted habitats (Wozencraft 1993). Carnivores do well in arid lands where they cope with life through behavioural, physiological, morphological and ecological adaptations (Wagner 1980; Cloudsley-Thompson 1993, 1996).

Arid lands occur extensively throughout the world. As extremely arid habitats, deserts are seemingly hostile to life. Yet they harbour a rich animal life (Schmidt-Nielsen 1964). In general, deserts are a relatively simple biome posing few problems for the organisms that inhabit it. As a result of climatic extremes, intraspecific competition for resources, and limited food resources, deserts have engendered the development of environmental adaptation extremes (Cloudsley-Thompson 1982). Nevertheless, in the desert biome the homogeneity of animal behaviour is far more apparent than its diversity (Cloudsley-Thompson 1993, 1996). The ability to obtain sufficient moisture for life through the body fluids of their prey has been a major key to the successful habitation of deserts and other arid areas by carnivores (Schmidt-Nielsen 1964).

An attempt to address the ecology of carnivores in arid lands adequately in all areas of the world is impossible in a single volume. This book therefore focuses on Africa and its adjacent countries, a region which has a rich variety of arid lands and carnivores. Nevertheless, the occasional reference is made to other regions to illustrate a specific point, or to serve as a contrast in adaptation. Even within Africa, comparisons are made between species living in arid and mesic environments to illustrate the degree and nature of carnivore adaptability. In this context, use is made of the Serengeti ecosystem and its adjacent areas in East Africa, as they represent a vast, natural, mesic but at times arid, system, where the wildlife and their life habits have been affected little by man.

Despite undergoing major geological changes, Africa has been one of the most environmentally stable continents in the world since the late Mesozoic. In the late Miocene, temporary land connections probably formed between

Africa and Eurasia, resulting in limited faunal interchange. As an example, the hyaenids came to Africa in that time. Major periods of faunal exchange occurred late in the Oligocene and Pliocene. From the Eocene to the late Miocene, most of Africa consisted of undulating plains seldom higher than 600 m above sea level, and its fauna altered little. Evergreen forests, savannas and arid areas persisted from the late Tertiary to the Miocene. Late in the Tertiary, major geologic upheavals created much of Africa's present relief, but change continued during the Pleistocene. The Great Rift Valley formed as recently as a million years ago. At least four glacial periods during the Pleistocene resulted at times in major climatic changes. These changes still continue. The last glacial period was some 70 000 years ago, and the last major cold spell 15 000 to 25 000 years ago. The most recent warm peak was 4000 to 8000 years ago. Protracted periods of low rainfall and resultant aridity have been common in Africa. Only 22 000 years ago, the southern fringes of the Sahara Desert were some 300 to 600 km south of where they are today (Delany and Happold 1979). These changes must have influenced the extent and severity of the arid zones of Africa, and consequently their wildlife.

Today there are three arid regions in Africa: the Sudanese, Somali and Southwest arid zones. The Sudanese arid zone lies north of 12°N and includes the Sahara Desert. The Somali arid zone comprises the coastal and near coastal regions of Ethiopia, Somalia, Kenya and northern Tanzania, and is connected with the Sudanese arid zone along the Red Sea coast. The Southwest arid zone includes the Namib Desert, and the Kalahari basin of South Africa, Namibia and Botswana. This zone was at times linked with the Somali arid zone through a drought corridor which narrowed and widened over time (Delany and Happold 1979). Today there is a mesic corridor of some 900 to 1 200 km wide separating these two zones (Skinner and Smithers 1990). Apart from the true deserts, there are various and extensive arid areas in Africa, such as the vast Karoo, Namaqualand and Bushmanland of South Africa, and the arid parts of northern Kenya and its adjacent countries. Even a mesic area such as the Serengeti ecosystem experiences extensive arid periods to which its fauna have to be adapted.

In general tems, extremely arid deserts receive 100 mm or less rain per year, arid deserts receive from 101 to 250 mm, and semi-arid lands from 251 to 500 mm. Yet rainfall alone does not define a desert because the total environment must be considered, including the various adaptations of its inhabitants (Lovegrove 1993). Most parts of the Namib Desert, for example, receive less than 100 mm of rain annually, yet it is a less extreme habitat than other deserts. Following years of more abundant rain, carnivores such as the bat-eared fox *Otocyon megalotis*, cheetah *Acinonyx jubatus* and yellow mongoose *Cynictis penicillata* may temporarily inhabit the Namib Desert, while the normally dry riverbeds, which run from the eastern highlands to the western coast, create favourable habitats for more permanent habitation by carnivores and other animals (Seely and Griffin 1986).

1.1
The Approach Used

The larger species are discussed at length because most of the known data about carnivores in arid regions concern them. Data from mesic regions are included for comparison. The herpestids are discussed in some detail to illustrate the adaptations found in smaller carnivores. The mustelids and viverrids are mentioned only briefly, however, due both to a lack of space and of relevant information. Where a specific species or subspecies is mentioned in the text for the first time, its scientific name is given. It may also be used again later, when necessary.

Carnivores from the following areas are discussed: the Kalahari ecosystem, a vast sandfilled basin wherein the rainfall increases from the southwest to the northeast (Mills 1990). This ecosystem occupies much of the centre of southern Africa. Even today, many parts of it are rarely visited by man (Main 1987). The southern Kalahari stretches across southeastern Namibia to northwestern South Africa and Botswana, and has a mean annual rainfall of some 220 to 240 mm/year (Mills 1983). It includes the Kalahari Gemsbok National Park of South Africa. The central Kalahari lies mainly in Botswana and receives a mean of some 250 to 450 mm of rain per year (Owens and Owens 1978). Nxai Pan in the northeastern Kalahari region of Botswana receives a mean of 500 to 550 mm. The whole Kalahari ecosystem experiences frequent, long, hot and dry periods (Main 1987). Further northwest in Botswana, the Moremi Wildlife Reserve lies on Kalahari sands along the edge of the Okavango Delta. It only receives some 450 to 500 mm of rain per year (Main 1987), while the Savuti Marsh of the northern Kalahari in the Chobe National Park in northeastern Botswana receives some 500 mm of rain per year (Cooper 1993). However, the water flowing into this marsh, which forms intermittantly in the Mababe Depression, comes and goes with minor geological upheavals, and the marsh may remain dry and disappear for many years before filling and forming again. It has recently been dry since 1981 (Main 1987). Consequently, it is currently a dry and arid "marsh".

Elsewhere in South Africa, the Postberg Private Nature Reserve and the adjacent West Coast National Park receive a mean of 270 mm of winter rain per year (Avenant 1993). Most parts of the Karoo region receive less than 500 mm: the Karoo National Park receives some 200 mm per year (Dean and Milton 1991), and the Mountain Zebra National Park 394 mm per year (Moolman 1989). The central karoid parts of the Cape Province of South Africa receive 437 mm of rain (Richardson 1987a,b). However, in its northwestern part, the Augrabies National Park receives less than 200 mm of rain per year (Bothma and Le Riche 1994a). The Bedford farming district in the Karoo receives 470 mm (Pringle and Pringle 1979).

The Suikerbosrand Nature Reserve, south of Johannesburg in South Africa, is a mesic area, which receives more than 700 mm of rain per year. The western, sandveld grasslands of central South Africa receive from 400 to 600 mm (Lynch 1980), and the eastern arid bushveld areas adjacent to the Kruger National Park, such as the Sabi Sand Game Reserve, and the Klaserie and Timbavati Private Nature Reserves, receive means of between 400 and 500 mm (Bearder 1977; Kruger 1988). The extensive Kruger National Park in northeastern South Africa receives a mean annual rainfall of 440 mm in the southwest to 740 mm in the northeast, but it experiences periodic, protracted droughts (Bailey 1993). The Limpopo River Valley in the Northern Province of South Africa gets a mean of 330 mm of rain per year (Berry 1981). South of the Kruger National Park, the Umfolozi Game Reserve receives more than 500 mm (Fuller and Darby 1958).

Further north in southern Africa, the Matobo National Park in Zimbabwe and the Kafue National Park in Zambia are relatively mesic, while the Etosha National Park in northern Namibia receives a mean of 351 mm of rain per year (Stander 1992a). The Namib Desert stretches along the entire western coast of Namibia, and although sparse, its rainfall increases inland from the coast. This coast, the Skeleton Coast, receives from 5 to 20 mm of rain per year. This increases to an erratic 15 to 110 mm, or occasionally even more, in the central Namib Desert (Lovegrove 1993).

The Serengeti ecosystem is a vast and austere region of plains and hills in northern Tanzania. It includes the Ngorongoro Crater on its southeastern edge, and the Serengeti Plains which stretch westward from the Ngorongoro Crater Highlands, but in their rain shadow (Schaller 1972; Hanby et al. 1995). The seasons differ considerably, and the rainfall varies from 700 to 800 mm of rain per year in the Serengeti National Park in the north. However, the eastern plains receive less rain. Moreover, extensive dry periods are experienced from time to time (Schaller 1972). Lake Manyara National Park lies southeast of the Serengeti ecosystem in the Rift Valley of Tanzania and receives a mean of 767 mm (Sinclair 1977). The Selous Game Reserve in southern Tanzania receives some 760 mm of rain per year (Rodgers 1974). The Masai Mara area, which includes parts of northern Tanzania and southwestern Kenya, receives 850 mm of rain per year in places (Malcolm 1986; Fuller and Kat 1993). However, the Rift Valley floor in much of Kenya is relatively arid (Fuller et al. 1988). The Masai Mara area includes the Masai Mara National Reserve in Kenya. The Tsavo National Park in southeastern Kenya receives variable rainfall. This is generally above 500 mm per year (Hamilton 1976), but the Taru Desert in southeastern Kenya lies in the Nyika rain shadow and receives only about 200 mm (Rasa, pers. com.[1]). The Nairobi National Park of Kenya is an area of relatively mesic, shortgrass plains (Eaton 1970a). Central Niger, an arid region of the Sahel, receives from 200 to 400 mm of rain per year (McShane and Grettenberger 1984), while the Bale Mountains of Ethiopia represent a high Afro-Alpine region. Although the rainfall is in excess of 1000 mm,

[1] Prof. O.A.E. Rasa, Arbeitsgruppe für Ethologie, University of Bonn, Germany.

the higher elevations are more moderate. The studies done there concerned the more arid, shortgrass and arid scrub plateaus and valleys (Sillero-Ziburi and Gottelli 1995).

Outside Africa , the central deserts of Oman cannot rely on rain every year (Lindsay and Macdonald 1986). The Judean Desert of Israel lies near the Dead Sea, where the rainfall ranges from 50 to 100 mm per year (Macdonald 1978). The Negev Desert south of the Dead Sea receives a mean of 90 mm (Bouskila 1984). Doñana National Park in Spain has a mediterranean climate with dry summers, receiving a mean of 529 mm of rain in winter (Beltràn and Delibes 1994). Korcula Island, which lies in the Adriatic Sea, also has dry summers, and receives 676 mm of rain each winter (Cavallini and Serafini 1995). The Gir Wildlife Sanctuary and Gir National Park lie in the Gir Forest of southwest India with 650 mm of rain per year in the east and 1000 mm in the west (Chellam and Johnsingh 1993). The Bandipur Tiger Reserve in India receives monsoon rains, but also experiences a long, hot and dry season (Johnsingh 1982). The Annapurna Mountain Range in Nepal is part of the Himalayas, but receives little precipitation: it is a rain shadow steppe (Oli et al. 1993). The arid area southwest of Alice Springs, Australia, where the feeding behaviour of the dingoes was studied, is a dry forest and scrub receiving only 173 mm of rain per year (Corbett and Newsome 1987). The other dingo study areas were more mesic (Newsome et al. 1983b).

CHAPTER 2

Family Canidae

2.1 Taxonomy, Distribution, and Status

The family Canidae occurs widely throughout the world, and has been successful in colonizing most of the land area of the earth. Its species occupy remarkably diverse ecological niches (Mech 1975). Through the dingo, *Canis lupus dingo*, the family has also colonized Australia. According to Wozencraft (1993), the specific epithets *familiaris* and *lupus* are synonymous. However, although *Canis lupus* and *Canis familiaris* were described simultaneously by Linnaeus in 1758, and *familiaris* has page preference over *lupus*, *Canis lupus* has become universally accepted for the species which includes the wolf, dingo and domestic dog.

No subfamilies are recognized for the Canidae, but 14 genera and 34 species are currently recognized (Wozencraft 1993). The habitats occupied are diverse, although few canids inhabit deep forests. Canids are primarily adapted for a cursorial life. They do not possess the morphological specializations necessary to enable them to bring down large prey as efficiently as felids, but they are also not as strictly carnivorous as felids (Kleiman and Eisenberg 1973).

From the listings of the Convention on International Trade in Endangered Species of Fauna and Flora (CITES) and the International Union for Conservation of Nature and Natural Resources (IUCN), it would appear that canids are less threatened in the world than felids. Nevertheless, of the canid species, 12 are in one or another Appendix of CITES. They are : *Canis aureus* of North and East Africa, southwest, central and southern Asia and southeastern Europe; *Canis lupus* which is found in most of the northern hemisphere; *Cerdocyon thous* of South America; *Chrysocyon brachyurus* of South America; *Cuon alpinus* of the East; *Pseudalopex culpaeus* of South America; *Pseudalopex griseus* of Chile and Argentina; *Pseudalopex gymnocercus* of South America; *Speothos venaticus* of North and Central America; *Vulpes bengalensis* of Asia; *Vulpes cana* of Asia, and *Vulpes vulpes* of which three Indian subspecies are listed in CITES, i.e. *Vulpes vulpes griffithi, Vulpes vulpes montana* and *Vulpes vulpes pusilla*. The following canid species are considered by the IUCN to be endangered: *Canis rufus* of the United States of America, *Canis simensis* of

Ethiopia and *Lycaon pictus* of Africa, while *Dusicyon australis* of the Falkland Islands is believed to be extinct (Wozencraft 1993).

Eleven canid species currently occur in Africa, of which *Canis adustus, Canis mesomelas, Canis simensis, Lycaon pictus, Otocyon megalotis, Vulpes chama* and *Vulpes pallida* only occur in Africa (Wozencraft 1993). Of these, the African wild dog *Lycaon pictus* has been persecuted most, and its range has recently declined severely. This species has been the victim of man's persecution, having been called murderous, an abomination, a blot upon other wild things and many other ill-considered and ignorant comments (Childes 1988). Previously abundant in sub-Saharan Africa, the African wild dog now only occurs in three substantial populations in conservation areas in northern Botswana (± 700 individuals), the Kruger National Park (350 to 400 individuals) and the Selous Game Reserve area (700 to 900 individuals) (Boggs and McNutt 1994).

It is interesting that the wolf is still said to inhabit parts of North Africa and that it is widely distributed, although not common, in the Arabian Peninsula (Wozencraft 1993; Nader 1996). Yet the wolf is by no means the largest carnivore to have been found in North Africa, as both the lion *Panthera leo*, and until recently, the bear *Ursus arctos* also occurred in northwest Africa (Wozencraft 1993). Of the social canids, the grey wolf and the African wild dog occupy similar niches, but they mainly occur at opposite ends of the world. Physically, the grey wolf and the African wild dog are similar, but the wolf has a much longer nose and smaller ears than the wild dog. The wolf and the African wild dog are direct ecological counterparts which have solved the same ecological problems in remarkably similar ways, despite living in widely different environments (Mech 1975). The African wild dog and the Asiatic wild dog or dhole *Cuon alpinus* are highly specialized descendants of the wolf-jackal group of canids (Grzimek 1975). The dog-like canids differ from the fox-like canids in having rounded pupils as opposed to elongated ones (Müller-Using 1975). Of all the canids, Blanford's fox *Vulpes cana* from the East is probably one of the rarest and prettiest. It is feline in appearance, with thick, soft fur, a long furry tail, large eyes and a round face. It is a mountain dweller in Asia, India and Russia where it lives on cliffs and in natural rock crevices. Its behaviour is also more feline than vulpine (Al-Khalili 1993).

2.2 Social Behaviour

Despite their diverse morphology and ecology, the social behaviour of the canids has remained remarkably similar throughout the family. This is especially evident in the postures and movements of the various species (Kleiman 1967). Canids are unusual among carnivorous mammals in that the members of a group share food and care for sick adults and dependent young (Moehlman 1989).

2.2.1 Group Formation

As body size increases in canids, there is a concurrent trend towards cooperative hunting. However, group size does not correspond strictly to prey size (Moehlman 1989). Specializations which have occurred in group-living canids serve to maintain group cohesion, and are of degree rather than kind between species. Nevertheless, the more solitary species are usually more fox-like, and the more social ones more dog-like. An exception, however, is the bat-eared fox which is more group oriented than would be expected for a fox (Kleiman 1967).

Wild Dogs. The African wild dog (Fig. 2.1) is intensely social. It lives in large packs in which a clear dominance hierarchy exists (Mech 1975; Grobler et al. 1984). It is also the only canid in which cooperative hunting may have been the major evolutionary force for the development of social behaviour (Moehlman 1989). The pack is a close feeding community and there is strict division of labour within it (Grzimek 1975). In the Ngorongoro Crater, a minimum pack size of four to six dogs seems to be necessary for effective functioning (Estes and Goddard 1967). Living in a group has two main benefits, reduced predation risk and improved foraging efficiency (Jennions and Macdonald 1994). Pack size varies from 17 to 43 individuals in southwestern Kenya (Fuller and Kat 1990).

Fig. 2.1. The African wild dog, an intensely social canid

The African wild dog pack is a discrete and cohesive social unit, which functions primarily as a hunting unit, and which cooperates closely. It has a regular leader (Estes and Goddard 1967). All its members behave tolerantly to each other and only the females on the Serengeti Plains occasionally have minor disagreements when they compete for caring of the pups. This happens because every female tries to nurse all the pups and to steal them for herself (Kühme 1965). The pack travels, hunts and rests together (Frame et al. 1979). An African wild dog pack differs from the usual pattern of mammalian social organization by being a stable group of genetically related males, while the females leave their natal pack to join other packs (Frame et al. 1979). However, in the Serengeti ecosystem, young males are recruited into their natal pack only when pup survival is low or when the male parent, or alpha male, dies (Burrows 1995). A pack in the Serengeti may also at times split into two. This process involves subordinate dogs of both sexes (Frame et al. 1979).

In the Moremi Wildlife Reserve, wild dog packs consist of extended family groups which are nomadic except when denning. Packs there vary from 2 to 50 individuals, depending upon the area and time of the year, but the mean pack size varies from seven to eight individuals. Larger packs usually occur where prey densities are high. Their initial size is dependent upon the sizes of each of two single-sex packs which join. Dispersal then influences pack size over time. Packs of less than four individuals are unusual, and most packs live there in a precarious balance between success and failure (Boggs and McNutt 1994).

The Asiatic wild dog of the Bandipur Tiger Reserve in India lives in an area which receives monsoon rains from April to August, but which is quite arid for part of the year. The mean pack size is 8.3 adults, but 16.0 when the pups are included. There are no other pack-hunting canids in the areas where these wild dogs occur (Johnsingh 1982).

Pack size in dingoes, which live in dry, sclerophyllous forests and woodland scrub in south-eastern Australia, is related inversely to the level of temporal instability in the environment. Frequent wildfires may prevent the staple, medium-sized prey and the supplementary large prey from being abundant, and the combination of these two classes of prey is thought to be necessary for a large pack size (Newsome et al. 1983b).

Jackals. The Bale Mountains in Ethiopia are home to the endemic and endangered Simien fox *Canis simensis*, which is also known as the Ethiopian wolf (Tyler 1975; Morris and Malcolm 1977; Wozencraft 1993; Sillero-Zubiri and Gottelli 1995). It occasionally forms small packs when hunting, but is not a wolf. Of all canids, it is closer to the jackals. The golden jackal *Canis aureus* of Africa, Asia and eastern Europe is closest of all canids to the wolf. The golden jackal also occasionally forms small packs (Wandry 1975), especially when it hunts. However, normally it occurs alone, in pairs or in small family groups. At favourable feeding sites, large groups do appear (Müller-Using 1975;

Khidas 1990). In Israel, golden jackals also form stable, social groups of up to 25 individuals around refuse dumps and feeding sites created for carnivores (Ferguson et al. 1983).

South of the Sahara, the side-striped jackal *Canis adustus* is widely distributed and lives singly, in pairs, small packs or family groups (Müller-Using 1975; Macdonald 1983; Smithers 1983). In the Southwest arid zone, the golden jackal is almost completely replaced by the black-backed jackal *Canis mesomelas*. The side-striped jackal is only found in the arid areas of northeastern South Africa, northwestern Namibia, northern Botswana and eastern Zimbabwe (Smithers 1983).

The black-backed jackal (Fig. 2.2) replaces the golden jackal completely in the arid regions of South Africa (Smithers 1983). Its social organization is similar to that of the golden jackal in the Serengeti and in Israel, where pairs are the usual social unit. However, groups of up to 60 black-backed jackals will feed close together in Namibia at seal rookeries along the Namib Desert coast, where dead pups form an abundant food resource, and elsewhere at carrion (Smithers 1971; Ferguson et al. 1983). Simien foxes usually hunt alone but may congregate at dawn or dusk (Morris and Malcolm 1977).

Foxes. Contrary to the general trend among canids to form smaller groups as body size decreases (Kleiman 1967), the bat-eared fox *Otocyon megalotis* (Fig. 2.3) does live in groups and is the most gregarious of all foxes (Nel and Bester 1983; Grobler et al. 1984). In the Masai Mara National Reserve, the

Fig. 2.2. The black-backed jackal which replaces the golden jackal in southern Africa

mean group size of the bat-eared fox is 2.8 individuals (Malcolm 1986). In the southern Kalahari, group size can vary dramatically over the years, with the mean fluctuating between 2.3 and 4.2 individuals. There is a linear increase in the frequency of occurrence of groups of bat-eared foxes larger than two as rainfall (and hence the prey base) increases in the southern Kalahari. Of these changes in the frequency of group size, 81% is due to rainfall. The duration of group cohesion in bat-eared foxes of the southern Kalahari is affected by food availability. Groups break up early during times of low prey availability (Nel et al. 1984). On the western, sandveld grasslands of central South Africa, group size of bat-eared foxes is also reduced in summer, from the time when family groups begin to forage together, to when they start mating and pupping. In winter, and after mating, the groups break up entirely (Mackie and Nel 1989).

2.2.2 Nomads

Individual nomads are rare among canids. However, African wild dog packs in the Moremi Wildlife Reserve are considered to be nomadic, except during the denning season (Boggs and McNutt 1994). In the Ngorongoro Crater, a pack of 21 wild dogs remained together for a period of 4 months, only appearing at irregular intervals at other times (Estes and Goddard 1967). Van Lawick and Van Lawick-Goodall (1970) also consider African wild dog packs in the arid southern part of the Serengeti ecosystem to be nomadic.

Fig. 2.3. Contrary to the general trend among smaller canids, the bat-eared fox lives in groups.

2.2.3 Communication

Canids generally have a wide range of similar postures and movements which are used for communication (Kleiman 1967). These will not, however, be dealt with in detail here. They also generally employ scent-marking and vocalization which increases in frequency in both sexes just before mating (Kleiman 1968).

As the largest living canid, the wolf uses a range of vocal, olfactory and other methods for communication (e.g. Asa et al. 1985; Paquet 1991). Many of these are also shown by other canids. The African wild dog is vocal when not hunting (Grobler et al. 1984), scent-marks frequently and also shows anal dragging, body rolling and body rubbing in communication (Van Heerden 1981). Scent-marking by African wild dogs is carried out by using urine and faeces in much the same way as in wolves (Mech 1975). The frequency of urine-marking is higher just before and during the breeding season, whereas marking with faeces is unaffected by breeding condition. In captive wolves, urine-marking is primarily directed to peripheral areas of the pen, and faeces to the more central ones (Asa et al. 1985).

The faeces of canids such as wolves, emit volatile compounds and can serve as more salient signals than urine. Alternatively, faeces may form a substrate for scent-marking with anal gland secretions (Asa et al. 1985). The same is true with other canids. Moreover, African wild dogs have a strong body odour. Sebaceous glands occur over the whole body and must leave scented messages when the dogs roll on the ground, rub themselves against objects or merely pass through an area. However, there is no tail gland in the African wild dog (Van Heerden 1981). Urine and faeces are deposited regularly in the same spot (Van Heerden 1981), but African wild dogs have apparently not yet been reported to make scrapes as wolves do (Paquet 1991). Only the dominant pair of a wild dog pack regularly urine-marks, often simultaneously and in the same place (Frame et al. 1979).

The Asiatic wild dog uses vocal communication, interdigital glands, scats and urine to scent-mark its range. Defecation sites are used as common latrines for faecal scent-marking (Johnsingh 1982), as is also commonly the case among hyaenas in the southern Kalahari (Mills 1990). Asiatic and African wild dogs which linger at a kill will follow the same route as that taken by other pack members, either by tracking interdigital gland secretions or by making use of traditional pathways (Johnsingh 1982).

The golden jackal of the Djurdjura National Park and the rural areas of Bangladesh has a complex social organization with clear seasonal rhythms (Khidas 1990; Jaeger et al. 1996). The golden jackal rings its territory with communal latrines (Macdonald 1983). Unlike the side-striped jackal, which does not give long, extended howls (Müller-Using 1975; Smithers 1983), the black-backed jackal is highly vocal (Müller-Using 1975). Black-backed jackals in the southern Kalahari scent-mark with urine and faeces (Ferguson et al. 1983), but social interactions are generally amicable. A fixed greeting ceremony takes place between pair members, and black-backed jackals have a well-developed social life (Ferguson 1978). Simien foxes in the Bale Mountains occasionally urine-mark on prominent objects (Morris and Malcolm 1977).

Field observations in the Serengeti ecosystem indicate that the bat-eared fox urinates and defecates while foraging, usually well within the home range boundaries (Lamprecht 1979). This fox uses its urine to scent-mark the boundaries of the home range, but defecation is not used in scent-marking (Smithers 1971; Lamprecht 1979). Bat-eared foxes in the southern Kalahari also defecate anywhere in the home range. Scent-marking with urine has only been noted in pairs or larger groups there, and frequently more so in winter. This is probably due to the process of pair formation, and not to mark food resources (Nel and Bester 1983).

As the most gregarious fox of all, the bat-eared fox in the southern Kalahari has a large repertoire of graded signals. Some of these have evolved through ritualization into displays which may have different meanings, depending upon the context within which they occur. Communication between bat-eared foxes has probably evolved in conjunction with their life-style in any given habitat. Environmental differences will therefore be reflected in the method and frequency of use of a particular channel of communication (Nel and Bester 1983).

Facial and tail postures are used by bat-eared foxes in the southern Kalahari to communicate, but some olfactory signals are also given. These foxes are not territorial and make few loud vocalizations. Vocal spacing mechanisms are not used at all. There is a higher incidence of calling in winter than in summer. This coincides with increased sexual activity but is not employed in territory establishment (Nel and Bester 1983). Bat-eared foxes in the southern Kalahari also call each other collectively to an abundant food source (Nel et al. 1984).

The fennec fox *Vulpes zerda* is the smallest of the canids that inhabits the deserts of North Africa and the Arabian Peninsula (Carlstead 1991). It has excellent hearing and uses a great variety of calls to communicate (Von Gauthier-Pilters 1967). It often digs connected dens in the sand (Müller-Using 1975). Most play activities in fennec foxes are related to instinctive behaviour (Von Gauthier-Pilters 1966). Blanford's fox, which occurs in the arid areas of the Arabian and Sinai Peninsulas, Palestine, Iran, Pakistan and India has, in general, more feline than canid (vulpine) characteristics (Al-Khalili 1993).

2.3 Reproduction

Reproductive behaviour in canids follows a general pattern in which breeding occurs seasonally, except in the case of the bush dog *Speothos venaticus* of the forests of northern South America and southern central America, where it seems to be all year round. In both canid sexes there is an increase in the frequency of scent-marking and vocalization during preoestrus, and increased agonistic encounters between males. Oestrus usually lasts for 2 to 5 days, with one to two copulations taking place per day. Gestation lasts for 50 to 70 days. The copulatory tie is a special feature of canid reproduction (Kleiman 1968). Large canid females also usually produce larger litters than small ones (Moehlman 1989).

The most common canid mating system, especially apparent in smaller (< 6 kg) forms like foxes, is obligatory monogamy, except on the rare occasions when the animals are polygynous. Such monogamy is rare among other mammals. The smaller canids generally also have a sex ratio favouring females. They have young female helpers and young males that disperse. Medium-sized (6 to 13 kg) canids like jackals and coyotes are also strictly monogamous, have equal sex ratios, female helpers and males that disperse. The larger canids like the African wild dog are also monogamous but they tend towards polyandry, with the sex ratio in adults favouring males. They have male helpers and females which disperse. Also, in canids generally, the mating system is greatly influenced by the size of the food used, and by the temporal and spatial availability of the food resource (Moehlman 1989).

2.3.1 Wild Dogs

In an African wild dog pack, one dominant or alpha pair does all the mating (Bertram 1979; Frame et al. 1979; Boggs and McNutt 1994; Burrows 1995). This pattern of limited reproduction has a great influence on the conservation of the species as it results in a small and fragile gene pool. In the Moremi Wildlife Reserve, for example, there are only 70 to 80 breeding pairs of wild dog in the total population of some 700. This happens because packs as large as 50 form when prey densities are high. Each of these packs only has one breeding pair. When prey densities are low in the dry season, the pack size is seven to eight (Boggs and McNutt 1994). In the less arid Serengeti ecosystem, there is high pup and alpha male mortality. This enables young males to breed in their natal packs, leading to high recruitment rates (Burrows 1995).

The African wild dog was formerly thought to be unique in not having a copulatory tie (Kleiman 1968), although it does have a *bulbus glandis*. However, it has now been shown that these dogs do have a copulatory tie, but that it does not last for more than 50 to 112 s (Van Heerden 1981). Relative to other canids, copulation in wild dogs is brief and usually lasts for no more than 5 min, followed by a gestation period of 60 to 73 days (Smithers 1983; Grobler et al. 1984). Litter size in African wild dogs varies. In the Serengeti ecosystem, a mean litter size of 10.1 (range 6 to 16) was recorded by Frame et al. (1979). This is regarded by Moehlman (1989) as generally true for most areas. The sex ratio of pups at birth is biased toward males (Malcolm and Marten 1982), as is revealed by a litter of eight males and one female recorded in the Ngorongoro Crater (Estes and Goddard 1967). In the arid bushveld of South Africa, the mean litter size on reaching 3 months of age is 11.9. Pup mortality between the ages of 4 and 10 months is severe (44%). Pup survival is also positively linked to the number of females in the pack (Maddock and Mills 1994). Although litter sizes at birth are likely to be determined genetically, pup survival is resource-related (Fanshawe et al. 1991). Hence aridity through its impact on such resources, may well play a role in pup survival after birth.

The pups of the African wild dog are small and dependent when born. This imposes severe pre- and post-partum investment costs on females, and helpers are essential to raise the pups (Moehlman 1989). In the Serengeti ecosystem, all the yearling and older members of the pack help to raise the pups (Frame et al. 1979; Burrows 1995), as they do in the Moremi Wildlife Reserve (Boggs and McNutt 1994). However, reliance on helpers can become a constraint on the overall survival of wild dogs. This may happen because its necessity effectively makes the energetic costs of breeding so high that a lone pair is incapable of reproducing successfully (Jennions and Macdonald 1994; Kühme 1965). Pups are suckled for 10 to 12 weeks (Kühme 1965; Grobler et al. 1984; Mills 1990), but they are also fed at the den with regurgitated food which they start to eat as young as 14 days old (Kühme 1965). When the wild dog pups on the Serengeti Plains are small, adults cache food for themselves near the den because they do not range widely then (Malcolm 1980). Three adults in a pack can regurgitate enough food to feed the pups when they kill prey regularly. When food becomes scarce, however, as in the dry season or in times of drought, the adult dogs continue to regurgitate food until it becomes necessary to abandon the pups. Yearlings and adults regurgitate comparable amounts of food in the Serengeti ecosystem. Yearlings also sometimes steal regurgitated food from the pups. Feeding the pups places no real strain on the pack as only a small portion of the food eaten is later regurgitated. Mothers protect their pups at the den for the first 6 weeks of life, when the rest of the pack goes off to hunt. This is much as Asiatic wild dog mothers do. Adult dhole males seldom stay with the pups (Johnsingh 1982; Malcolm an Marten 1982). The den of the Asiatic wild dog has a play and rest area 2 to 3 m away from it (Johnsingh 1982).

2.3.2 Jackals

The golden jackal is monogamous, but the pair bond is not as strong as in the black-backed jackal, and mate changes do happen (Moehlman 1989). During the breeding season, the male follows the female constantly and she often snaps at him (Müller-Using 1975). In the Serengeti ecosystem, the Ngorongoro Crater and the Rift Valley of Kenya, the breeding season of the golden jackal is in summer (Wyman 1967; Müller-Using 1975; Fuller et al. 1989), but in Bangladesh pups are born year-round (Poché et al. 1987). The difference is possibly due to environmental influences and the use of another food resource. Gestation lasts for 62 to 63 days (Müller-Using 1975) and the mean litter size in the Serengeti ecosystem is 2 (range one to three) (Wyman 1967). The male provisions his mate during pregnancy (Moehlman 1989). Both parents help to raise the pups, and they also regurgitate food for their young (Wyman 1967). The male is particularly aggressive towards strangers during this time (Wandry 1975). Of the pups dispersing in a given year, 70% will return to their parents to help feed the lactating female and the pups of the next year. Pup survival does not seem to be limited by food provisioning. However, young golden jackals which leave their natal areas, do so in the dry season food bottleneck (Moehlman 1989).

The monogamous black-backed jackal is one of a few mammal species which has a long-lasting pair bond (Wyman 1967; Smithers 1983; Moehlman 1989). This reflects the physiological and ecological constraints of its environment (Moehlman 1989). In southern Africa, young black-backed jackals are born in winter and early spring (Wyman 1967; Bothma 1971a), as they are in the Rift Valley (Fuller et al. 1989), but the breeding season varies with local conditions. Fertility is generally high (Müller-Using 1975; De Vos 1970; Grobler et al. 1984). Pseudopregnancy, which lasts almost as long (58 days) as normal gestation (60 to 70 days), can occur. Such pseudopregnancy is probably due to spontaneous ovulation. At the end of pseudopregnancy the female shows all the behaviour normally associated with preparatory parturition (De Vos 1970).

Some known mean litter sizes in the black-backed jackal are 5.7 (range one to nine) (Grobler et al. 1984; Moehlman 1989) and 4.2 (range two to six) (Bothma 1971a). Parental care is critical to pup survival. The female cannot raise pups on her own. Raising the pups also requires a substantial male investment (Moehlman 1989). Parental care of the young by the pair is about equal, but the male also feeds the lactating female (Moehlman 1979). When such a male parent dies, whole litters may be lost and the female may disappear. The litter is totally dependent upon its parents for at least the first 4 months of life, but the parents will allow helpers to assist in raising their young (Moehlman 1989). These helpers are the young of the previous year that remain with, or return to their parents, the following year. They bring food to the lactating female and her pups and teach the pups to groom and hunt. They also guard the pups when the parents are away hunting (Moehlman 1979; Smithers 1983). Their presence at a den correlates directly with increased pup survival and is independent of food density. In return the helpers themselves benefit from the extended experience gained in the natal territory (Moehlman 1979).

On the Serengeti Plains, 73% of black-backed jackal litters have helpers. However, a pair with a first litter cannot have helpers as it has had no previous young. Black-backed jackal pups first emerge from the den when about three weeks old (Moehlman 1979). By this time all the deciduous teeth have errupted (Lombaard 1971). The pups are weaned when eight to nine weeks old (Bothma 1971a; Moehlman 1979). Most mortality occurs in the first 14 weeks of life (Moehlman 1979). In the southern Kalahari and elsewhere, the pups move from the close proximity of the den when 3 months old, but they stay in the general vicinity for at least 6 months, finally dispersing when about 2 years old (Ferguson et al. 1983).

In its habits, the side-striped jackal differs greatly from golden and black-backed jackals (Müller-Using 1975). However, its reproduction follows the established canid pattern. Breeding usually occurs in late winter to spring just before the first rains in southern Africa, although some births have been recorded in summer. Gestation lasts for 57 to 70 days (De Vos 1970; Grobler et al. 1984), and a mean litter of 5.4 (range four to six) is born in an underground den usually located in an old aardvark *Orycteropus afer* burrow (Smithers 1983; Grobler et al. 1984). Of the Simien fox little is known, but it does seem to live in pairs too (Müller-Using 1975). It may well exhibit the same general reproductive characteristics of other canids.

2.3.3 Foxes

The Cape fox *Vulpes chama* has a litter of three to five, which are born in summer after 50 days of gestation. The pups eat meat when still quite young (Grobler et al. 1984). The bat-eared fox is monogamous, but compared with other canids it produces small litters. Its pups are more developed at birth. In the Serengeti ecosystem, the first pups are usually born from September to October after a gestation period of 60 to 65 days (Lamprecht 1979; Grobler et al. 1984; Malcolm 1986). The mean litter size is 3.2 and the two nearest pupping dens found were 300 m apart (Malcolm 1986). The young first emerge from the natal den when 17 days old (Smithers 1971). All members of a group care in some way for the pups (Malcolm 1986). The pups suckle until they are about 15 weeks old. This is longer than in most other canids (Lamprecht 1979). The parents carry small prey items to the pups, but they do not regurgitate food. The male guards the pups while the female forages (Malcolm 1986), and initiates the pups into foraging (Nel 1978). When the pups are moved to new dens, they are carried there by both the parents (Lamprecht 1979).

In southern Africa, the pups of the bat-eared fox are born later than in East Africa. In the Free State Province of South Africa and in the southern Kalahari, the pups are born in summer from October to late January, but mostly in November (Nel et al. 1984; Mackie and Nel 1989). Bat-eared foxes remain monogamous there for several years, and the pair bond is renewed in July after which the pair will return to a den site used the previous year (Mackie and Nel 1989). In the southern Kalahari, pair bonds in bat-eared foxes last for long periods and these foxes first breed when 18 months old. Rainfall, and hence aridity, possibly influences either litter size, pup survival, or both through its influence on food availability (Nel et al. 1984). The female suckles the pups before foraging, and returns to do so again after 9 to 10 h of foraging (Nel 1978).

Rüppell's fox *Vulpes rüppelli* in the central deserts of Oman also forms monogamous pairs which share a den (Lindsay and Macdonald 1986). However, no other data on its reproduction are known. The fennec fox, which inhabits the deserts of North Africa, and the Arabian and Sinai Peninsulas, gives birth to a litter of two to five young born from February to March in an underground den. This den is often connected under the sand with other dens nearby. The gestation period varies from 49 to 52 days (Von Gauthier-Pilters 1967; Müller-Using 1975; Walker 1983). The pups are weaned when they are 61 to 70 days old. Security at the den is most important since the female is most sensitive to den disturbance just after the young have been born. The male defends the female before and during the birth of the young (Von Koenig 1970).

2.4
Area Use, Movements, and Activity

Variations in food size, and the temporal and spatial availability of food, affect canid spacing systems greatly. Hence these must be influenced by increased aridity and its ecological consequences. Ecological constraints among territorial canids determine whether it is possible for additional individuals to remain with the parental pair (Moehlman 1989). This must be especially true of the jackals which usually form pair bonds.

2.4.1
Wild Dogs

Throughout its former distribution in sub-Saharan Africa, the African wild dog has declined. In many places it has disappeared entirely because of its large range requirements and conflict with humans. Of all surviving canids, this has led to it being the most endangered. Relict populations with little or no chance of survival are possibly still found in a few countries including Algeria and Senegal, but the only countries which currently contain potentially viable populations are in eastern and southern Africa, where most of the world's estimated population of 3000 to 5000 wild dogs are found today (Fanshawe et al. 1991; Fig. 2.4). Even in conservation areas, huge game reserves or national parks are necessary for the survival of this large, social carnivore because its minimum viable population in a given area is in excess of 25 individuals (Childes 1988). Few protected areas in Africa today are large enough for safe populations of 200 to 300 animals (Fanshawe et al. 1991).

As nomadic, social animals, the territory of a pack of African wild dogs varies greatly in size from 500 to 1500 km², with ranges generally overlapping by 50 to 80%. In East Africa, the Serengeti ecosystem is home to a fairly large proportion of the world's African wild dog population. All the wild dogs in the Serengeti form one breeding population and they are not genetically isolated. However, the distribution of these dogs is patchy, the preferred habitat being the grassland plains with an occasional occurrence in the open woodlands. Pack range estimates in the Serengeti vary from 50 to 260 km² (Burrows 1995), but can be as large as 2000 km² (Moehlman 1989). Contrary to what Kühme (1965) believes, these ranges are not exclusive (Burrows 1995) .The size of the range of a given pack at any particular time is related to the abundance of food in the area (Kühme 1965). Hence it must also change with increased aridity. When food becomes scarce, or when denning has been completed, wild dogs may shift their range. A pack may therefore range over an area of 1500 to 2000 km² in a given year while sharing this space with three or four other packs. The total range size of the African wild dog in southwestern Kenya is relatively small (660 km²) compared with those in other areas of East Africa. This is mainly because of the year-round presence of abundant prey (Fuller and Kat 1990). The movements of a pack are also related to hunting success. When prey is scarce, such as during the dry season, the whole range will be traversed in 2 to 3 days. However, when there are pups at the natal den, the pack's movements are more restricted. The dominant pair usually determines the movements of the pack, either by direct leadership or by refusing to follow subordinate individuals (Frame et al. 1979).

Only the Hangwe and Zambezi areas of Zimbabwe are large enough to ensure the long-term survival of African wild dogs in that country, where they are today confined to safari areas and national parks. Even so, the existing total population does not exceed 350 individuals (Childes 1988). Elsewhere in southern Africa, there are a few remnant populations of the African wild dog in Botswana, South Africa, and perhaps a few other localities. The population in the Kruger National Park and its neighbouring private nature reserves, a conservation area of some 21 350 km², has been stable at some 357 dogs over the past 25 years. The stability is due to stable prey populations and a low incidence of disease and poaching. The minimum density there is one

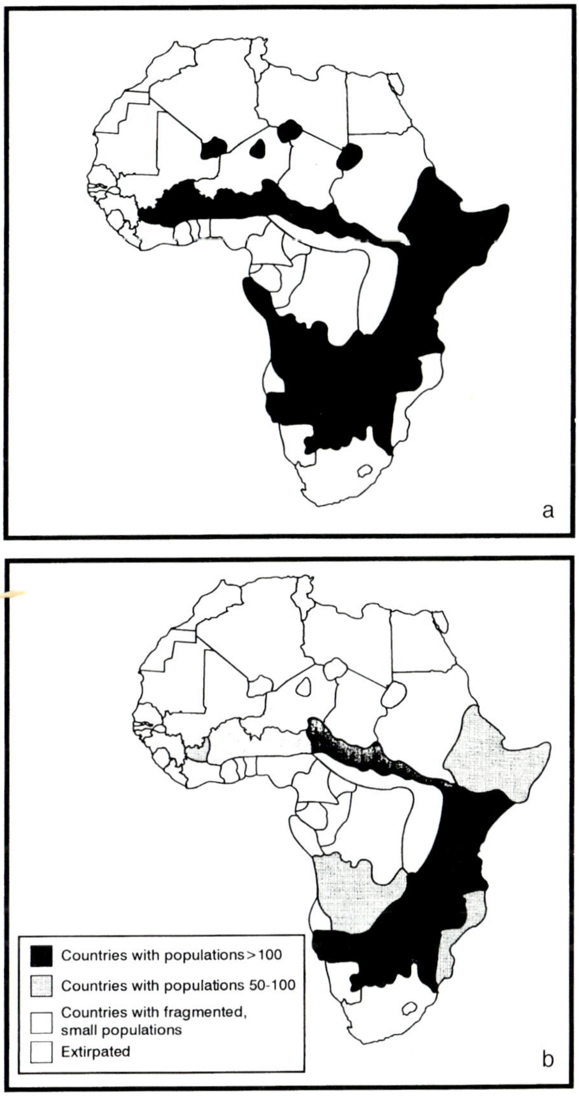

Fig. 2.4 a, b. Distribution of the African wild dog in **a** 1980 and **b** 1988 to show the recent decline in its range (Fanshawe et al. 1991)

dog per 60 km² (Maddock and Mills 1994). In the Kruger National Park, wild dog ranges are smaller (500 km²) than in the Serengeti (1500 to 2000 km²). This may be because of the migratory herds present in the Serengeti. The wild dog density in the Kruger National Park is also higher than in the Serengeti (Moehlman 1989). In the Moremi Wildlife Reserve, wild dog packs range over areas of some 450 km², one-third of which is shared with neighbouring packs (Boggs and McNutt 1994).

African wild dogs hunt in daytime, usually just after sunrise or before sunset, but not at night (Kühme 1965; Boggs and McNutt 1994). In the Moremi Wildlife Reserve, wild dogs travel a mean distance of 10 km daily, but occasionally as much as 20 km in a morning. Male subadults there disperse when they are 18 to 20 months old, and females when 30 to 32 months old (Boggs and McNutt 1994). The dispersal of the sexes of the African wild dog in the Moremi Wildlife Reserve is the opposite to that found in the sexes in East Africa. The males of Moremi may disperse over distances of up to 150 km, while the females establish themselves adjacent to their natal range, as do most other mammals (Boggs and McNutt1994).

Asiatic wild dogs in the Bandipur Tiger Reserve have small total pack ranges of 40 km^2, with core areas of 20 km^2. Relative to those of African wild dogs, these are small. The difference is in part due to the inherent sedentariness of Asiatic wild dogs, and the high prey biomass available to them. As with African wild dogs, the size of the pack decreases when the dogs are denning. The mean density varies from 1.1 to 2.9 animals per km^2 (Johnsingh 1982), which is considerably higher than that of African wild dogs.

2.4.2 Jackals

The golden jackal uses thickets and the dens of other animals for shelter (Müller-Using 1975). It usually has a small territory of less than 1 km^2. The territories on the open, shortgrass Serengeti Plains form a light mosaic (Moehlman 1989). In the Serengeti, however, the territories have a diameter of about 3.2 km, varying with local topography. The territory is defended by complex displays (Wyman 1967). In an arid part of the Rift Valley in Kenya, the mean size of a golden jackal's territory is 2.4 km^2 (Fuller et al. 1989).

In contrast, an abundant food resource at a garbage dump leads to an extremely small mean territory size of 0.1 km^2 for golden jackals in one area of Israel. However, even such small territories are marked on their boundaries by middens (Macdonald 1983; Moehlman 1989). Golden jackals in the Djurdjura National Park also have a complex territorial system. Here the territory boundaries are marked, although the territories may in part overlap. Territories are especially important during the breeding season (Khidas 1990). Golden jackal territories in Bangladesh are 1.1 km^2 for males and 0.6 km^2 for females (Poché et al. 1987). These figures are closer to the norm of less than 1.0 km^2 specified by Moehlman (1989). Based upon its behavioural attributes, Wandry (1975) believes that golden jackals can live at higher densities than those in which they are currently found. In a part of the Rift Valley, the golden jackal is the only one which is active by day, but not during the middle of the night (Fuller et al. 1989).

A pair of black-backed jackals is exclusively territorial, and the male and female mark their territory (Nel 1978; Smithers 1983). The size of the territory depends upon the food available (Grobler et al. 1984). On the Serengeti Plains, the territories of black-backed jackals also form a mosaic, as with golden jackals, but they have core-areas of use in the centre. Boundary fights sometimes occur between animals of the same sex (Moehlman 1979). In the Rift Valley, black-backed jackals have rather small mean home ranges of 1.8 km², probably because of an unexpectedly abundant food resource (Fuller et al. 1989).

The black-backed jackal is abundant in southern Africa where it has adapted successfully to life in many arid areas. On the grassland plains, the territory of an adult black-backed jackal pair is mutually exclusive. It varies greatly from 1.3 to 575 km² (mean: 50.0 km²) between habitats. However, the mean range for young jackals is 28.9 km², for subadults it is 85.2 km² and for adult pairs 10.6 km² (Ferguson et al. 1983). This region is often highly agricultural. Consequently, prey resources must be relatively low.

In the southern Kalahari, where there is usually an abundant food resource of small mammals, an adult pair of black-backed jackals has a mean territory size of 4.2 km² from which both sexes chase away intruders. Disputes often lead to fights between the pairs. Nevertheless, the territory of a pair may overlap that of neighbouring pairs by up to 10%. Since not all adult animals are mated and have fixed ranges, some movement by unmated adults across the ranges of more than one pair may occur. Although exclusively territorial, territory defence breaks down in black-backed jackals of the southern Kalahari at abundant food such as an antelope carcass, and at a scarce resource such as a waterhole. When this happens, the intruders show submissive behaviour to the resident pair. Black-backed jackal pairs also tolerate young and immature animals within their range as long as the latter show clear submission (Nel 1978; Ferguson et al. 1983).

The Namib Desert is even more arid than the southern Kalahari. The range of a black-backed jackal pair there varies, but the minimum area is 24.9 km². Such ranges are not considered to be territories, because of the need for range overlap due to large, but localized food resources (cormorant and seal rookeries) and the otherwise homogeneous habitat. However, prey abundance rather than habitat structure and prey distribution is the major factor influencing range size in the Namib Desert (Hiscocks and Perrin 1988).

The black-backed jackal moves around extensively, and is active both by day and at night. Especially in areas where there are few people and little disturbance, such as in national parks, it is quite active during the day (Smithers 1983; Bothma et al. 1984; Hiscocks and Perrin 1988). In the Suikerbosrand Nature Reserve, adult jackals follow an activity pattern which is closely related to that of their main prey (mainly murids). This pattern is, however, influenced by light intensity. The lowest activity levels occur in bright sunshine, and on moonless nights when the vision of the jackals is impaired. In the southern Kalahari, the initiation and cessation of activity in winter and summer are directly related to the times of sunset and sunrise. Along with the amount of available moonlight, sunset and sunrise determine the precise time of onset of activity of a given pair of black-backed jackals on any specific night. This is ultimately dependent upon prey activity as is mentioned above. Innate patterns of activity (the endogenous circadian rhythm), on the other hand, are determined by ecological factors, such as the relative humidity of the air, temperature and food supply (Ferguson et al. 1988).

The Skeleton Coast of Namibia is cold and windy. Along this coast, the black-backed jackal has to forage on the beaches where it finds its major food resource. However, it lies up in coastal vegetation hummocks to save energy (Dreyer and Nel 1990). Black-backed jackals normally move around considerably. In one study on the northwestern grasslands of South Africa, five adult ear-tagged jackals travelled a mean distance of 13 km away from their point of capture over various periods of time (Bothma 1971b). In the southern Kalahari, a pair of jackals moved a mean of 10.6 km in 24 h in summer, and 7.8 km in winter. Adult jackals (daily mean: 11.8 km) move further than younger ones (daily mean: 8.8 km) (Ferguson et al. 1988).

Dispersing, young black-backed jackals may travel considerable distances away from their natal dens. On the western, sandveld grasslands of South Africa, pups 4 to 5 months old are still in their natal area, but one young male who was ear-tagged when 8 months old, was found 103 km in a straight line away from the point of capture 8 months later (Bothma 1971b). Another young male, almost 2 years old when radio-collared, moved 112.6 km in a straight line away from the point of capture in 15 months. It is clear that while some black-backed jackals act as helpers and remain close to their natal den, others range widely (Ferguson et al. 1983), probably when they are still single.

Prey density and predictability are the major factors which influence black-backed jackal density on the Skeleton Coast (Nel and Loutit 1986). A density of 22 animals per km^2 as is found around a seal colony is atypically high for black-backed jackals. However, it happens because of the presence of abundant food (Hiscocks and Perrin 1988). Elsewhere along the coast it is lower (Dreyer and Nel 1990).

A side-striped jackal pair in an arid part of the Rift Valley occupies a mean territory of at least 1.1 km^2. This is considered to be so small due to the paradoxically abundant food resource which is available to those jackals. The main activity periods are near sunrise and sunset (Fuller et al. 1989). The side-striped jackal is generally considered to be nocturnal (Müller-Using 1975). When moving, a side-striped jackal walks along and does not trot as a black-backed jackal does (Smithers 1983). Although the Simien fox is territorial and primarily active by day, little is known about its range-use and movements (Müller-Using 1975; Morris and Malcolm 1977). In some areas, however, it has become nocturnal, probably as a result of disturbance by humans. It does not burrow, but it may use caves and rock crevices for shelter (Morris and Malcolm 1977). The Simien fox of the Bale Mountains shows no evidence of territoriality and the hunting ranges of different individuals overlap extensively (Morris and Malcolm 1977).

2.4.3
Foxes

Bat-eared foxes of the Serengeti National Park are mainly nocturnal, with 85% of all activity occurring after dark. They may, however, forage in daylight on cool afternoons or in winter, as they do in the Namib Desert and elsewhere (Rautenbach and Nel 1978; Lamprecht 1979; Bothma et al. 1984). Their range in the Serengeti varies from 0.25 to 1.5 km^2 and, contrary to the arid southern Kalahari, overlaps in range are minimal (Lamprecht 1979). The mean range of a bat-eared fox in the Masai Mara National Reserve is 3.53 km^2. This is larger than in the Serengeti or Namib, but different ranges overlap as extensively (Malcolm 1986) as they do elsewhere (e.g. Nel 1978; Mackie and Nel 1989). In the Serengeti, bat-eared foxes only become resident in areas where there are sufficient patches of grass burnt short or grazed so by ungulates (Lamprecht 1979). When the migrating wildebeest *Connochaetes taurinus* arrive on the Serengeti Plains, the ideal shortgrass conditions which they create by grazing, make it possible for bat-eared foxes to reach a density of up to six foxes per km^2 in localized areas (Malcolm 1986).

The ranges of adult bat-eared foxes on the western, sandveld grasslands of South Africa overlap and vary from 0.82 km^2 in winter to 1.28 km^2 in summer. When suckling young, the females forage alone, while the males and pups remain in small areas measuring 0.37 to 0.47 km^2. The selection of range size, and hence of habitat, and the spatial and temporal use patterns of the range are all related to the relative presence of harvester termites. These termites are an important food for bat-eared foxes. This is in turn linked to aridity and the height of the vegetation as the foxes clearly prefer areas where the grass is 100 to 125 mm tall (Mackie and Nel 1989).

In the southern Kalahari, with its sparse grasslands, the bat-eared fox is not territorial (Nel 1978; Nel and Bester 1983) and ranges overlap extensively. The foxes there have a foraging range of 1 to 2 km^2. In the cool to cold winters the foxes are chiefly diurnal, but in the extremely hot summers they become nocturnal (Nel 1978). In contrast, the bat-eared foxes of Nxai Pan in the northeastern Kalahari are said to be active during the day despite the high temperatures at times. Regrettably, the season when the study was carried out is unknown. Although bat-eared foxes may also be nocturnal there, their main activity period is closely related to the availability of harvester termites which are maximally active and abundant on the open pan surface around noon. These termites are the main food for the bat-eared foxes there. As elsewhere in the southern Kalahari, the foxes are not territorial and intolerance between pairs has not been observed (Koop and Velimirov 1982). In winter, the bat-eared foxes of the Namib Desert emerge relatively early in the afternoon, at times just after noon, to start foraging, and 84% of all activity occurs before midnight. In summer, activity only starts late in the afternoon (Bothma et al. 1984).

Cape foxes have mutually exclusive ranges in the southern Kalahari (Nel 1978). They are strictly nocturnal and are only rarely seen by day. When they are seen, it is only late in the afternoon (Bothma 1966; Bothma et al. 1984). Rüppell's fox, which lives in the central deserts of Oman, rests by day in a den which is moved once every 4.8 days. The area used by a monogamous pair by day does not overlap with that of other pairs; but the mean range of 30.4 km² used at night does overlap between different pairs. The mean range of a given pair is 69.1 km² (Lindsay and Macdonald 1986). In the same desert, the red fox *Vulpes vulpes* rests by day in sheltered spots above ground, and its dens are shifted almost daily.The total range of one radio-tagged individual was 170.9 km² and the denning range was 5.5 km² (Lindsay and Macdonald 1986).

The small fennec fox is an agile and vigilant animal well-adapted to life in a severe environment. It is nocturnal and lives in small groups in the deserts of North Africa and the Sinai and Arabian Peninsulas (Von Gauthier-Pilters 1967; Müller-Using 1975; Walker 1983; Carlstead 1991). In summer, but not in winter, it usually lives in burrows up to 1 m under the sand (Von Gauthier-Pilters 1967; Walker 1983). The fennec fox can dig rapidly (Walker 1983), and this, together with its nocturnal habit, allow it to escape the intense heat of the day (Maloiy et al. 1982).

2.5 Feeding Ecology

Canids are specialized primarily to run down prey. They cannot bring large prey down as quickly and efficiently as felids do, nor when hunting do they use vegetation for cover so effectively. Small prey are usually killed by holding them in the jaws, then crushing and breaking the neck with vigorous shakes. The pack hunters kill their prey by repeated biting. The victim eventually weakens through exhaustion and injury, whereupon it is killed by evisceration or bites to other body parts (Kleiman and Eisenberg 1973). Canids are typically lithe and muscular runners that are capable of speeds of up to 30 km/h for extended periods (Moehlman 1989). All are opportunistic hunters and their adaptability and non-specialization enables them to survive in a great variety of environments (Kleiman 1967). The smaller canids tend to be solitary hunters, the medium-sized ones hunt cooperatively at times, while for the large ones cooperative hunting is important, but not critical in obtaining food (Moehlman 1989).

Pack hunting is uncommon in the family. It is most obligatory and ritualized in the African wild dog where it plays an important role in maintaining social cohesion within the pack (Kleiman 1967; Moehlman 1989). However, as body size increases, there is an increased trend toward some degree of cooperative hunting (Moehlman 1989). As with other carnivores, the length of time spent pursuing a given prey is variable and adjustable

(Anholt et al. 1987). The canids are not as strictly carnivorous as the felids. This makes them greater opportunists and contributes towards peaceful communal feeding, which is rare among felids (Kleiman and Eisenberg 1973).

The type and size of its food tend to correlate with a specific canid's spacing and feeding system. Small species depending on invertebrates and fruit for food, have overlapping ranges and little impact on each other's foraging success. Among the territorial and solitary feeders, the distribution of food may determine territory and group size. Where populations use food that varies seasonally or annually, their territory size may be determined by the minimal food resource level (Moehlman 1989).

2.5.1 Food

Wolves. The only known data on the largest canid known from truly arid areas, the Arabian wolf *Canis lupus arabs*, concerns its food selection. Ever since its contact with man, the Arabian wolf has been known to prey upon domestic sheep and goats, but it also eats gazelles, carrion and even garbage. However, this wolf has remarkable adaptability and variability of diet. In extremely arid areas where large prey is rare, wolves feed on various small prey including invertebrates (Nader 1996). Arabian wolves living near a permanent stream in Wadi Turabak in the Sarawat Mountain Range of arid southwestern Saudi Arabia feed on the sacred baboon *Papio hamadryas*, the European quail *Coturnix coturnix*, an endemic fish *Cyprinion malensis* (Cyprinidae) and even some unidentified freshwater snails. In a semi-desert area of Iran the wolf *Canis lupus pallipes* kills gazelles, dogs, hares, rodents and partridges (Biquard et al. 1994). In the Judean Desert, the wolf also scavenges, and although it does not eat dogs, it does hunt gazelles *Gazella dorcas* (Skinner and Ilani 1979).

Wild Dogs. The African wild dog is the only canid which uses large prey as food throughout the year (Moehlman 1989). In the Serengeti ecosystem, wild dogs scavenge a portion of their food (Bertram 1979) as they do in Kenya and elsewhere (Fuller and Kat 1990). Scavenging requires the ability to travel long distances with low energy expenditure (Bertram 1979). Serengeti wild dogs kill Burchell's zebra *Equus burchellii* mares more often than stallions because the mares tire more quickly (Bertram 1979). They also take more male than female Thomson's gazelles *Gazella thomsonii* due to shorter flight distances in the males. The preference for male gazelles is a consequence of territoriality in the gazelles. The male gazelles are usually in poorer condition than the females on account of breeding stress and being more prone to sarcoptic mange than females. It is however, not their poor condition which leads to their selection as food, but other ecological circumstances. There is still no clear evidence that the wild dogs of the Serengeti select old or sick animals (FitzGibbon and Fanshawe 1989).

Of the prey taken by African wild dogs in the Ngorongoro Crater, the majority (54%; n = 50) are Thomson's gazelles, with newborn and juvenile wildebeest second (36%) (Estes and Goddard 1967). In the Masai Mara area, wild dogs also mostly kill Thomson's gazelles and young wildebeest (Fuller et al. 1995). With a more varied and different prey base available to them, wild dogs in the Kafue National Park take 15 types of prey, with the common duiker *Sylvicapra grimmia* (26%) and reedbuck *Redunca arundinum* (25%) being eaten most. The habitat of the reedbuck is largely inundated during the wet season there, affording it temporary refuge from predation. Consequently, it is more often killed by wild dogs during the dry season (Mitchell et al. 1965). In the Kruger National Park, wild dogs show a clear preference (93%) for impala *Aepyceros melampus*, but they will kill a wide range of other prey, including the occasional honey-badger *Mellivora capensis* or young lion *Panthera leo* (Grobler et al. 1984). However, they rarely take carrion.

African wild dogs in the Moremi Wildlife Reserve also mainly eat impalas, which form 85% of the larger prey available there. However, these wild dogs also eat reedbuck, lechwe *Kobus leche*, kudu *Tragelaphus strepsiceros*, tsessebe *Damaliscus lunatus*, bushbuck *Tragelaphus scriptus*, springbok *Antidorcas marsupialis* and steenbok *Raphicerus campestris* (Boggs and McNutt 1994). In the agricultural areas of Namibia, African wild dogs prey on domestic livestock (Hines 1990), but in the Etosha National Park they eat the abundant springbok and Burchell's zebra, and even springhares *Pedetes capensis* (Scheepers and Venzke 1995). The African wild dog occasionally uses other carnivores as food. In Kenya, they chase and even eat lions (Fuller and Kat 1990). Lions are also killed and eaten in the Kafue National Park (Mitchell et al. 1965).

In comparison, the Australian dingo *Canis lupus dingo* mainly eats medium- sized and large mammals, but it changes its diet in accordance with the relative abundance of its prey, especially during drought (Newsome et al. 1983a). Dingoes only eat rabbits *Oryctolagus cuniculus* consistently, but they will devour rodents and even lizards at times. Large prey are only eaten in large amounts during drought. Then the carcasses of red kangaroos *Macropus rufus* and cattle which have died of the drought are scavenged (Corbett and Newsome 1987).

Jackals. Golden jackals live in a varied range of envionmental conditions where they may experience seasonal and annual fluctuations in food (Moehlman 1989). In the Serengeti ecosystem, they eat a variety of food from plant material to invertebrates and vertebrates (Müller-Using 1975). Although they also scavenge, carrion forms a minor part of their diet (Houston 1979). In the Serengeti National Park and the Ngorongoro Crater, golden jackals are important predators of Thomson's gazelles, especially lambs. However, these gazelles are less numerous in the Crater than on the Serengeti Plains. Consequently the golden jackals in the Crater scavenge a fair portion of their food, although not nearly to the same extent as black-backed jackals do there (Wyman 1967).

In stark contrast, golden jackals living in an arid region of central Niger mainly eat vegetable matter (grasses, seeds and fruits: 83% occurrence in all scats; n = 143) and invertebrates (71% of all scats). In addition, they eat the rodents *Gerbillus* sp., *Acomys* sp. and *Xerus* sp. (62%), hares *Lepus* sp. (31%), birds (24%), small gazelles *Gazella* sp. (4% of all scats) and a few reptiles. In the cold, dry season when natural food is scarce, they also turn to domestic sheep and goats as an alternative food source (McShane and Grettenberger 1984). In Bangladesh, on the other hand, the golden jackal seldom feeds on field rodents but relies heavily (45% of all stomachs examined) on food scavenged in and around towns (Poché et al. 1987).

Black-backed jackals in the Serengeti are important predators of Thomson's gazelle lambs, but in the Ngorongoro Crater they scavenge much (35%) of their food (Wyman 1967). In southern Africa, the black-backed jackal's diet is equally diverse. It is omnivorous, although it mainly eats vertebrates (n = 425 stomachs) (Bothma 1971c; Smithers 1983). Plant material is also eaten, but not to a major extent. Wild fruits, and grass seeds and leaves are usually the material taken (Bothma 1971c).

The most interesting dietary adaptations of the black-backed jackal occur along the Skeleton Coast of Namibia. In the central part of the Namib Desert, the diet of the black-backed jackal is as varied as elsewhere (Bothma et al. 1984). Plant material was found in 92% of all the scats (n = 87) in one area, and plays an important role in the diet of these jackals. In particular the fruits of *Euclea pseudebenus* are eaten, and up to 177 seeds of this plant have been found in a single jackal scat. Elsewhere in the central Namib, where rodents are the most abundant food, the same *Euclea* sp. fruits are again an abundant item in all scats (83%; n = 772), while some invertebrates are also eaten (Stuart1976). However, along the Skeleton Coast the food resources of carnivores are clumped (seal and bird rookeries) and variable. In some areas birds are the predominant food (93% of all scats) of black-backed jackals. In others, plant material, and especially the large prickly fruits of *Aconthosicyos horridis*, are eaten predominantly. The skin of this fruit is first peeled and the jackals then eat its seeds and flesh (Stuart 1976). Elsewhere, Cape fur seals *Arctocephalus pusillus* are the main food of jackals (n = 33 scats). They are probably mainly scavenged (Stuart and Shaughnessy 1984). Black-backed jackals are known to wade up to their bellies into the surf to retrieve seal pup carcasses at several seal colonies. They also consume seal faeces which is rich in fish remains, and even eat the remains of mussels *Perna perna* dropped onto rocks by kelp gulls *Larus dominicanus*.

The seal colonies of this harsh desert coast provide an abundant and predictable food resource throughout the year for black-backed jackals and other carnivores such as the brown hyaena *Parahyaena brunnea*. Jackals regularly kill seal pups too (n = 47 scats) (Hiscocks and Perrin 1987). Elsewhere along the coast, the washed-up carcasses of fishes and other marine animals are eaten. The jackals are even known to enter temporary lagoons of the major rivers after flooding to catch live mullet fish *Mugil cephalus* trapped in the shallows by the receding water (n = 234 scats) (Nel and Loutit 1986). They also eat the sea catfish *Tachysurus feliceps*, but the method of obtaining it is unknown. Such fish may possibly have been regurgitated by Cape cormorants *Phalacrocorax capensis* or white-breasted cormorants *Phalacrocorax carbo* that had been disturbed while feeding (n = 312 midden samples) (Avery et al. 1987). Jackal scats picked up as far as 18 km inland, contain food of coastal origin. This is a clear indication of the importance of this food resource for the survival of both local and adjacent black-backed jackal populations (Dreyer and Nel 1990).

The side-striped jackal is also omnivorous. In Zimbabwe, its diet often (72% of all scats) includes plant material, with wild figs *Ficus* sp. and wild prunes *Parinari curatellifolia* eaten most often. However, murids (35%), hares and invertebrates are readily eaten too. They less often eat carrion, birds and reptiles (Smithers 1983). In South Africa too, wild fruits are an important source of food for side-striped jackals, but there the principal fruits eaten are *Diospyros* sp. and *Ziziphus* sp. (n = 15 stomachs) (Bothma 1971d). Müller-Using (1975) confirms that side-striped jackals supplement their diet substantially with plant material. The side-striped jackal does not scavenge as much as the black-backed jackal does (Grobler et al. 1984).

According to Müller-Using (1975), the Simien fox lives almost exclusively on rodents (Sillero-Ziburi and Gottelli 1995). This animal is a distinctive and very localized endemic canid of Ethiopia. It is most abundant in the moorlands of the Bale Mountains, which support a rich array of rodent populations. This abundance is reflected in its diet. The diet includes the giant mole-rat *Tachyoryctes macrocephalus* (42% of its prey by weight), and other rodents and hares of the alpine moors. The giant mole-rat sometimes emerges briefly on the open, alpine moorland to forage, and is then caught. Despite also being locally abundant, shrews are not eaten at all by the Simien fox, probably because their large skin glands make them distasteful (Morris and Malcolm 1977). The Simien fox also scavenges its food occasionally, especially the prey of eagles. The Simien fox is an integral component of Ethiopia's high-altitude ecological balance, and is one of the few canids that thrive in open country at altitudes above 3000 m elevation (Morris and Malcolm 1977; Sillero-Ziburi and Gottelli 1995). However, and ironically so, it is precisely this refined adaptation to an abundant Afro-Alpine rodent prey base which today constrains these foxes to a scarce and fragmented habitat. This may eventually lead to their extinction from ecological changes initiated by human activities (Sillero-Ziburi and Gottelli 1995).

Foxes. Of the true foxes, the Cape fox is also omnivorous, but in its preferred habitat of semi-arid, sandy plains or scrubland, it mainly relies on rodents for food (Bothma 1966, 1971d). Surprisingly, and unlike the Simien fox, which is not a true fox, the diet of the Cape fox also includes shrews (Bothma 1971d; Grobler et al. 1984). Murids are its main prey (88% of all scats) in the Namib Desert, with hares and insects playing a supplementary role (n = 25 scats) (Bothma et al. 1984). Rüppell's fox of the central deserts of Oman, is equally omnivorous, mainly eating small mammals, birds, lizards and insects, a diet similar to that of the red fox in the same area (Lindsay and Macdonald 1986). Blanford's fox, one of the rarest and prettiest foxes of Asia, is a mountain dweller. It lives on cliffs and in rocky crevices, and is mainly insectivorous in diet (Al-Khalili 1993).

The predominantly soft diet of termites and other insects (Bothma 1971d) of the bat-eared fox is not reflected in an increase of shearing facets on its molars as might be expected. Instead, its dental development is robust (Kieser 1995), possibly reflecting its canid origins and allowing for a diverse diet if necessary . Bat-eared foxes in the Serengeti ecosystem eat insects mainly, but also the eggs and nestlings of ground-nesting birds (Lamprecht 1979). In the Masai Mara National Reserve, bat-eared foxes also mainly eat insects (Malcolm 1986). In Botswana the diet is similar , but there bat-eared foxes may also eat droppings of springbok which are full of termites at times (Smithers 1971). In northern South Africa, along the arid Limpopo River Valley, their diet of insects (57% by volume) is generously supplemented by wild fruits (33%), especially those of *Grewia* sp. and *Arachis* sp. Of the insects eaten, 78% by volume are harvester termites *Hodotermes mossambicus*. These foxes also occasionally eat birds (n = 18 stomachs) (Berry 1981) as they do in the Serengeti, but they rarely eat carrion (Grobler et al. 1984).

Invertebrates are also the main food of bat-eared foxes in the southern Kalahari, with harvester termites predominating. However, food selection remains opportunistic, with *Grewia* sp. berries being eaten extensively (50% by volume) when available in summer. In winter, on the other hand, ants rather than termites, form the bulk of their food (n = 382 scats) (Nel 1978). When food becomes scarce during drought in the normally dry riverbeds, the primary habitat of bat-eared foxes in the Kalahari Gemsbok National Park, the foxes move into the interior dunes where the survival of rodents and insectivores is possibly greater. These dunes may therefore act as an ultimate habitat for the survival of bat-eared foxes in that arid region (Nel et al. 1984). On Nxai Pan, the presence of foraging termites on the pan surface makes it unnecessary for the foxes to dig for their food as is generally believed to be the case (Koop an Velimirov 1982). The bat-eared foxes of the Namib Desert prey mainly on the termite *Hodotermes mossambicus* as they do elsewhere, but they also feed on the long-nosed harvester termite *Trinervitermes trinervoides* and other insects (n = 23 scats). There is no correlation between food selection by bat-eared foxes and rainfall in the Namib Desert as might be expected (Bothma et al. 1984).

The fennec fox is mainly carnivorous, feeding on lizards, birds, bird's eggs, small rodents, insects (especially migratory locusts) and even plants (Von Gauthier-Pilters 1967; Maloiy et al. 1982; Walker 1983).

2.5.2 Hunting and Foraging

Wolves. The Arabian wolf hunts singly, or in pairs. It also occasionally hunts in small groups (Nader 1996).

Wild Dogs. As a cursorial pack hunter, the African wild dog gives the impression of being a tireless runner (Walker 1983). However, most hunts are not overly long (Mech 1975). The African wild dog mainly uses its vision to hunt, but stragglers during a chase may use smell to come back on track. A potential prey group is located visually, and the pack then openly moves slowly closer over a wide front to approach the prey. When close, all the dogs charge the herd. As soon as the herd starts to flee, the individual quarry is selected visually by one of the pack leaders. The chase can last several kilometres at speeds of up to 66 km/h. Once the chase is on, other potential prey are ignored. Any pack member which catches up with the prey, especially when it is large, tears out lumps of muscle or the intestines whilst it is still running, thereby weakening it progressively. The hunt ends when the prey is either exhausted or weakened sufficiently to falter or stop: it is then killed (Estes and Goddard 1967; Grzimek 1975; Bertram 1979; Smithers 1983; Grobler et al. 1984).

In the Serengeti ecosystem, where at times prey is relatively numerous, African wild dogs hunt only in the early morning or late afternoon, and never at night. The rest of the day is spent in and around the dens (Grzimek 1975). There, 88% of all hunts take place in the short to medium open grass plains. Elsewhere in Africa, wild dogs hunt equally successfully in closed bush and woodland (Fuller and Kat 1990). When the wildebeest migration passes through their area, the wild dogs hunt over a limited range of some 52 km^2. When these move away, however, the pack may hunt over a range of 300 to 400 km^2 or even more, traversing it every 2 to 3 days (Kühme 1965; Frame et al. 1979).

Adults 2 to 6 years old do most of the hunting. Yearlings may also take part in the hunt but they hesitate to tackle large prey (Malcolm and Marten 1982). Kühme (1965) has never seen a relay hunt and Smithers (1983) believes that one of the pack leaders always remains in the lead. Estes and Goddard (1967) also state that wild dogs do not hunt in relays as is commonly believed, but that flanking dogs stay behind the leader to intercept the prey if it circles or dodges back. When a pack member gets lost during a hunt, it calls repeatedly with a bell-like sound, then raises its head to listen attentively. The entire pack usually comes running over within a few minutes (Grzimek 1975).

Wild dogs have high hunting success rates in the Serengeti. These rates are related to hunting technique to a much greater extent than in the other large carnivores present there (Bertram 1979). Nevertheless, hunting by wild dogs does not disrupt the prey populations because these dogs are highly selective in the Serengeti, specializing mainly on the abundant Thomson's gazelles and young wildebeest for food (Estes and Goddard 1967).

The wild dogs of the Masai Mara area behave differently from those in the Serengeti National Park in that they hunt at any hour of the day or night (Fuller et al.1995). One pack studied there was no less successful at hunting when it was abruptly reduced from 29 to 19 members by the death of four older members and the dispersal of six young dogs (Fuller and Kat 1993).

As in the Serengeti, the wild dogs of the Moremi Wildlife Reserve hunt in the early morning or late evening. Contrary to what occurs elsewhere, however, a wild dog pack there divides to chase several prey, regrouping at the starting point to learn if a kill has been made. Any dog that kills, at once returns to the starting point to collect the other pack members. The wild dogs of Moremi therefore do not hunt as single packs, although they do hunt in smaller groups. Moreover, a single, older and more experienced dog may make the kill as it is impossible for a pack leader to be present in all hunting groups (Boggs and McNutt 1994). Experience is essential for hunting success. This was demonstrated by the failure to reintroduce African wild dogs into the Etosha National Park, probably because none of the animals released were fully adult or experienced hunters (Hines 1990).

Jackals. The golden jackal is believed by Moehlman (1989) to usually hunt cooperatively, with several individuals moving about together. When a single prey is spotted, however, a single jackal and not the whole group will attempt to catch it (Müller-Using 1975). The black-backed jackal trots along through its range when hunting, using it superb vision and excellent sense of smell and hearing to locate food (Müller-Using 1975; Smithers 1983). In the southern Kalahari, hunting by black-backed jackals is not confined to specific times of the day or night. Seasonally, more hunts take place in summer because insect and rodent prey are abundant then (Ferguson et al. 1988). In the central Namib Desert, 22% of foraging black-backed jackals hunt in pairs; otherwise they hunt alone (Bothma et al. 1984).

The Simien fox feeds on rodents which it catches on the run (Müller-Using 1975). While these foxes were formerly believed not to dig up rodents (Müller-Using 1975), it is now known that they do at times dig rodents from burrows. The Simien fox takes short, tense steps in a crouched position when hunting, freezing in mid-step for up to 10 min. It occasionally pounces on prey in tall grass. The ability to freeze helps it to hunt giant mole-rats successfully when these emerge on open moorland (Sillero-Zubiri and Gottelli 1995). Prey are sometimes stalked for more than 30 min. The striking consistency of the prey throughout the year may be explained by the ability of the foxes to locate and kill the same species during every season. This is in contrast to other rodent-eating canids in which the seasonality of rodent prey abundance has a major influence on diet (Sillero-Zubiri and Gottelli 1995). At the same time, the Simien fox has retained the basic canid ability to hunt cooperatively, which it does at certain times (Sillero-Zubiri and Gottelli 1995). This contradicts the opinion expressed by Morris and Malcolm (1977) that the Simien fox never hunts cooperatively. The hunting ranges of individuals overlap extensively and Simien foxes may congregate at dawn and dusk (Sillero-Zubiri and Gottelli 1995).

Foxes. Foraging bat-eared foxes have little impact on each other's success (Moehlman 1989). Food is located mainly by sound while the fox walks slowly along. Smaller prey are caught on the ground surface, or are dug out (Smithers 1983; Malcolm 1986). The occasional rodent is also run down (Walker 1983). The pounce, which is common in other canids, is rare among bat-eared foxes (Nel 1978). In the Serengeti National Park, bat-eared foxes forage mainly at night, but they may also do so on cool afternoons. Communal foraging by family units is common. This functions as anti-predator behaviour. It results in the optimal exploitation of available food in a time-dependent way, and helps in locating patchy sources of insect prey (Lamprecht 1979; Malcolm 1986). When foraging, a bat-eared fox will hold its head low, with the ears pointed forward at an angle of 45°. The ears invariably flip back when prey has been captured, but as soon as chewing starts, they move forward again (Nel 1978).

The presence of termites on the surface of Nxai Pan around noon, makes it unnecessary for the bat-eared foxes there to dig for prey. This shows that bat-eared foxes can modify their foraging behaviour to exploit an abundant and easily accessible food resource. Even when accessible only in daytime, such a food resource is energetically inexpensive to exploit, especially in arid regions (Koop and Velimirov 1982). The fennec fox forages as other foxes do, but it also digs for food (Müller-Using 1975; Walker 1983).

2.5.3 Killing

Canids and hyaenids display a high density of features in the microwear of their canine teeth which reflects the similarity of their killing techniques: using a strong bite (Anyonge 1995). In southwestern Kenya, the kill rate of different African wild dog packs varies according to pack size, prey size and abundance (Fuller and Kat 1990). In the Ngorongoro Crater, where prey is abundant, the kill rate exceeds 85% (Estes and Goddard 1967). In the Serengeti ecosystem, it is as high as 89% (Schaller 1972). However, in the Masai Mara area, the overall hunting success of wild dogs is only 51%, although some species are hunted more successfully than others (Fuller and Kat 1993; Fuller et al. 1995). A pack of wild dogs in the Moremi Wildlife Reserve kills two medium-sized antelope such as an impala every three days (Boggs and McNutt 1994). In Kenya, some wild dog packs acquire the habit of killing Burchell's zebras mainly (Fuller and Kat 1993).

Surplus killing, a relatively rare phenomenon in carnivores, has not been reported among canids in arid regions. Surplus killing involves the killing at one time of a large number of prey animals by a predator, or by a pack of them, but not for the purpose of eating the prey, although they would normally do so. These killings are the consequence of behavioural compromises in predator and prey to meet opposing environmental requirements (Kruuk 1972a). It is assumed to be rare in arid areas where large numbers of prey seldom congregate.

2.5.4 Feeding

Carnivores of different taxa possess teeth adapted for specific functions and for specific types of food (Van Valkenburgh 1996). Canids rely more heavily on their canines and incisors for feeding than felids do (Anyonge 1996). Felids for example eat little bone, but the African wild dog and spotted hyaena eat bone and/or muscle in 20 to 30 % of all cases. They both also use the front paws to aid in feeding in more than 50% of all feeding bouts. Wild dogs chew food much more often than lions, cheetahs or spotted hyaenas do, except when feeding on skin. Generally, there is a negative relationship between the size of a predator and the need to chew, but the long muzzle of the wild dog has a mechanical advantage for chewing over other larger predators (Van Valkenburgh 1996).

The dominant pair in a pack of African wild dogs has feeding priority in the Serengeti (Frame et al. 1979). In the Masai Mara area, wild dogs consume all the edible portions of their prey. Their mean daily consumption of meat varies from 80 to 200 g of meat per kg of dog per day (Fuller et al. 1995), or 1.2 to 5.9 kg of food per adult dog per day. Of this, about 1 kg per dog per day is regurgitated for the pups after every full meal. These consumption rates are similar to those of other pack-hunting carnivores (Fuller and Kat 1990).

The scene at an African wild dog kill in the Moremi Wildlife Reserve is quiet and orderly, notably unlike that at a lion kill. There is a hierarchy of access to the food which is generally opposite to that of most other large carnivores. Any pups present get top priority, followed in sequence by yearlings, the dominant pair, other adults and subordinates. A large litter of young pups may even eat all the food at a kill, leaving nothing for the other pack members (Boggs an McNutt 1994).

It was previously thought that the African wild dog never cached food, and that it was one of a few canids which did not do so. One reason for this belief was that their large range makes it improbable for them to visit a cache site regularly to retrieve the food (Macdonald 1976). However, on the Serengeti Plains, adult male and female wild dogs do occasionally cache food. Nevertheless, caching is unusual. On all known occasions, these caches were made when the adults returned to the breeding dens after feeding. A food cache may or may not be covered with vegetation and soil. Caching only seems to occur when the pups are small, and the wild dogs do not range as widely as they normally will do (Macdonald 1976; Malcolm 1980). Such conditions might occur in most prey-poor regions. Despite the highly cooperative hunting and feeding of young, individual wild dogs do not share cached food with other members of the pack (Malcolm 1980).

The golden jackal defends its kills fiercely, possibly because it is more strictly carnivorous than other jackals (Moehlman 1989). A kill is rarely eaten at the kill site, but is usually dragged off to some cover for feeding, as in most other wild canids (Müller-Using 1975). Golden jackals and black-backed jackals cache food for later use. Such food is usually buried to evade scavengers. When scavenging, golden jackals of the Ngorongoro Crater keep away from the kills of other predators, but black-backed jackals may feed with spotted hyaenas on a kill. When spotted hyaenas are not present at carrion, golden and black-backed jackals will feed side by side (Wyman 1967).

Black-backed jackals in irregularly spaced and localized areas along the Skeleton Coast live in an unique environment with high onshore winds, low ambient temperatures and low overall food availability. There they carry whole or parts of their food to specific feeding sites where great accumulations of debris develop in well-defined middens, especially where jackal numbers are low. This trait is unique to these coastal jackals and is only found in areas with vegetation hummocks. The vegetation of the feeding sites on top of these hummocks offer protection against the cold winds which commonly occur there, especially in the morning. The use of these hummocks as feeding sites is probably related to energy conservation because the jackals living along this coast forage mostly from the late afternoon through the night when the ambient temperature can be low (Dreyer and Nel 1990).

In an equally unique environment, the Simien fox of the Bale Mountains digs holes in the ground where it caches food, which is then covered with soil and vegetation (Sillero-Zubiri and Gottelli 1995). In the southern Kalahari, bat-eared foxes apparently do not cache food (Nel 1978), but in the Sahara Desert and elsewhere the fennec fox does this in holes covered with sand, and prey is carried to shelter away from the killing site before being eaten (Von Gauthier-Pilters1967).

2.5.5
Water Use and Adaptive Physiology

Terrestrial carnivores are generally independent of free water, but drink water when it is available. Many canids live on pre-formed water, and the water obtained through metabolism. This, however, requires more prey than that needed to provide energy alone. In arid areas, water is clearly correlated with metabolism, and markedly related to body size in canids such as Rüppell's fox, which lives in a severe desert area (Golightly and Ohmart 1984). The African wild dog drinks water regularly, as does the black-backed jackal when available, although the jackal is independent of water (Grobler et al.1984). In the Namib Desert, where dense fog occurs routinely, black-backed jackals lick condensed fog from rocks and vegetation (Hiscocks and Perrin 1987).

The fennec fox, which lives in dry, inhospitable areas of the Sahara, Sinai and Arabian deserts, can subsist without free water for long periods, but will drink freely when water is available (Walker 1983). It is the only Saharan carnivore that is totally independent of drinking water (Von Gauthier-Pilters 1967). It has a low metabolic rate and low water expenditure, both of which are general features of most truly desert mammals. Low metabolic rates help to conserve water which would otherwise be wasted by evaporative cooling. Panting only begins at ambient temperatures of 35 to 37 °C. This indicates that evaporative cooling becomes necessary at times. When panting, the rate can be as high as 690 breaths per minute. This rate considerably exceeds that of other mammals of similar body size. When evaporative cooling does become necessary, therefore, the fennec fox is highly efficient (Maloiy et al. 1982).

The fennec fox has a combination of other physiological and behavioural features which allows it to adapt successfully to its harsh environment. It is nocturnal and reduces its need of water for heat regulation by resting in a burrow by day (Von Gauthier-Pilters 1967). It also has a low heart rate , and it has adjustable "heat gates" in the ears and feet for dissipating heat by non-evaporative means (Maloiy et al. 1982). Its thick covering of fur is atypical of desert animals. This is probably not an adaptation to daytime heat, but rather to the cold nights experienced in its environment (Müller-Using 1975). The intense heat of the day is ameliorated by its small body size and burrowing habits. The fennec fox can withdraw to an equable microhabitat during the intense, daytime heat stress period of the desert. As in other carnivores, the overall thermal conductance of the fennec fox is low. This enables it to maintain a body temperature of 38 °C even at an ambient temperature as low as 7 °C. At ambient temperatures above 35 °C, the fennec fox is equally efficient thermally, as it can increase its own body temperature to as much as 40.9 °C. This happens through an increase in the rate of panting at a metabolic cost, and by an increased circulatory adjustment to heat (Maloiy et al. 1982).

An elevated body temperature in a hot desert has some obvious advantages. These include the ability to store heat, and is a determinant of the direction and magnitude of heat fluxes between the fennec fox and its environment. Panting is the principal mechanism for evaporative cooling and for avoiding hyperthermia. The fennec fox can lose 55 to 60% of its total metabolic heat by this means. Once the ambient temperature approaches body temperature, an abrupt increase in thermal conductance occurs by the dilation of the blood vessels in the legs and large ears. The ears of the fennec fox form 16 to 20% of the body surface and are relatively the largest among foxes, often being 150 mm long or even more (Maloiy et al. 1982; Walker 1983).

2.6 Population Ecology

The major causes of population decline of the African wild dog are interrelated. They are: loss of large areas of habitat, persecution by man, diseases and competition (especially from lions and spotted hyaenas *Crocuta crocuta*) (Fanshawe et al. 1991; Burrows 1995). In the name of predator control, the African wild dog has already suffered a near fatal population decline in Africa (Boggs and McNutt 1994). In the Etosha National Park, separate attempts in 1978, 1989 and 1990 to reintroduce African wild dogs all failed because the dogs were released into sub-optimal conditions in areas of the park where prey densities are low, and where the dogs were vulnerable to predation by lions (Scheepers and Venzke 1995).

The pups of an African wild dog are especially susceptable to mortality and 90% of them never reach maturity (Fanshawe et al. 1991). On the Serengeti Plains, lions repeatedly visit the den sites of wild dogs, causing severe disturbance and threat. Following one such incident, the mother of four young cubs carried each of them singly in her mouth to a new den over 1 km away (Kühme 1965). In many areas of the shortgrass Serengeti Plains, an impermeable carbonate hardpan occurs about 1 m below the surface. This hardpan forms the floor of many wild dog dens and upon which the pups lie. Such dens are often damp or wet in the wet season, and hypothermia and drowning due to flooding kills some pups (Burrows 1995). In the Moremi Wildlife Reserve, the mean mortality rate of adult wild dogs is 30 to 40%, and that of the pups 50 to 60%, although the latter fluctuates greatly (Boggs and McNutt 1994). In the Kruger National Park, disease has not yet been proved to be an important cause of death in wild dogs. Tooth fractures are also infrequent. Nevertheless there is a high population turnover rate of wild dogs in this park (Van Heerden et al. 1995). It is known, however, that the African wild dog is one of the potential reservoirs of the parasite *Babesia canis* in South Africa (Van Heerden 1980).

Private game reserves and the Kruger National Park form a large contiguous conservation area of some 21 350 km² in the arid eastern bushveld areas of South Africa. Prey populations are stable and there is a low incidence of diseases in, and poaching of, African wild dogs. As a consequence, this area provides an environment conducive to wild dog survival. The wild dog population in the area has been stable at around 357 individuals over the past 25 years. The sex ratio of the dogs is close to parity, and at least 69% of the packs raise young in any given year (Maddock and Mills 1994).

One means of population regulation in the Asiatic wild dog is the restriction of breeding to one female. An increase in population size beyond a given point leads to restlessness and emigration. Mortality is also a major population regulator (Johnsingh 1982).

Not much is known about other aspects of the population ecology of jackals, especially in arid areas. However, on the coast of the Namib Desert, prey density and predictability seem to be major factors regulating jackal populations (Nel and Loutit 1986). In South Africa, the black-backed jackal, like the African wild dog, is a potential reservoir of *Babesia canis* (Van Heerden 1980).

Even less is known of the smaller canids in arid areas. The endemic Simien fox of the Bale Mountains has a low population of 200 to 250 individuals and is under severe threat due to human activities (Sillero-Zubiri and Gottelli 1995). In South Africa, bat-eared foxes of the central grasslands are considered to be an important regulator of harvester termites (Mackie and Nel 1989). In the southern Kalahari, the bat-eared fox is an opportunistic strategist whose population numbers depend upon the available food, and therefore fluctuates according to rainfall, and hence aridity (Nel et al. 1984).

2.7 Interaction with Other Carnivores

2.7.1 Wild Dogs

The different sizes of the predators in the Serengeti, their habitat choices, and various hunting methods all result in relatively low levels of competition. However, spotted hyaenas drive wild dogs from a considerable proportion of their prey (Bertram 1979). For the African wild dog, little data on interaction with other predators is available from arid areas, but they may occasionally kill a young lion or honey-badger (Grobler et al. 1984). However, in the Ngorongoro Crater, the spotted hyaena is a major competitor of the wild dog, being capable of driving small packs from their prey (Estes and Goddard1967).

In contrast, spotted hyaenas in southwestern Kenya are absent from 59% of all wild dog kills, and when present, are usually attacked and chased away by the dogs. No wild dog kills are appropiated by spotted hyaenas when prey is abundant, although they are scavenged after the dogs have left. When prey is abundant, the spotted hyaenas harass the dogs less often than when it is scarce. In times of scarcity, spotted hyaenas may have a major impact on wild dog populations by appropriating kills and by direct killing of pups. Wild dogs in this region chase and even eat lions, but lions more commonly steal wild dog kills or even kill the dogs (Fuller and Kat 1990; Fuller et al. 1995). In the Masai Mara area, tawny eagles *Aquila rapax*, vultures, black-backed jackals and spotted hyaenas are common at wild dog kills, although the eagles and jackals do not take over a wild dog kill. Smaller packs of wild dog there lose a significant portion of their kills to competitors, but this does not lead to nutritional stress (Fuller et al. 1995). In the Kruger National Park, wild dogs avoid habitats frequented by lions and spotted hyaenas (Grobler et al. 1984; Van Heerden et al. 1995).

2.7.2 Jackals

From January to April, golden and black-backed jackals occur in pairs and they also occur together on the Serengeti Plains where they have similar diets. Yet, they seldom compete because of the abundance of Thomson's gazelle lambs which form some 50% of their food. However, from August to December the black-backed jackal occurs in pairs more often than the golden jackal, which tends to be solitary then. The black-backed jackal then takes many more Thomson's gazelles than does the golden jackal. However, on the edge of the Serengeti Plains, the diets of golden and black-backed jackals are remarkably different from August to December, and the potential for exploitative competition is reduced. On the whole, the black-backed jackal seems to prefer the woodland edge areas of the Serengeti Plains, whereas golden jackals prefer open plains, although they also occur on the edge of the Plains. This spatial separation largely eliminates competition betweem them (Wyman 1967; Senzota 1990).

In both the Serengeti and the Ngorongoro Crater, spotted hyaenas sometimes raid black-backed jackal dens, killing the pups (Moehlman 1989). At hyaena kills in the Ngorongoro Crater, black-backed jackals often feed with the spotted hyaenas at a kill, while golden jackals stay some distance away. However, when the hyaenas leave, the two species of jackal will feed side by side on the remains of the carcass (Wyman 1967).

An arid part of the Rift Valley is home to golden, black-backed and side-striped jackals. Although sympatric, these jackals are separated ecologically, and this reduces the potential for competition. Of the three, the golden jackal prefers well-drained, open, grassy hilltops, the black-backed jackal lives mainly in light *Acacia* spp. scrub and woodland, and the side-striped jackal lives in the dense undergrowth along the watercourses which dissect the plateaus (Fuller et al. 1989). In southwestern Asia, the golden jackal only has competition from the striped hyaena (Müller-Using 1975).

In South Africa, a region which is generally more arid than East Africa, the side-striped jackal is displaced by the black-backed jackal in all but the extreme northern areas, where the two species are sympatric (Fuller et al. 1989). In the southern Kalahari, black-backed jackals are often preyed upon by leopards, but the jackals in turn will prey upon the young of lions, leopards *Panthera pardus* (Eloff 1973a; Bothma and Le Riche 1984, 1986, 1994a) and probably other carnivores too.

2.7.3
Foxes

The central Namib Desert is a harsh environment with unpredictable rainfall. It sympatrically harbours the aardwolf *Proteles cristatus*, black-backed jackal, Cape fox and bat-eared fox, all of which potentially compete for limited food resources. However, sufficient separation exists in the diet, use of space and activity patterns to reduce competition and to make co-existence possible. There is some overlap in the consumption of insects by the bat-eared fox, the black-backed jackal and aardwolf. Between the aardwolf and the bat-eared fox there is direct exploitative competition as both species only eat insects. The Cape fox also eats some insects, but it has a varied diet which also includes rodents. However, there is clear ecological separation in diet between the aardwolf and the bat-eared fox in terms of the choice of insects by species. The aardwolf almost exclusively eats long-nosed harvester termites *Trinervitermes trinervoides*, while the bat-eared fox relies heavily on another species of harvester termite, *Hodotermes mossambicus*. This selection separation is quite stable and neither the aardwolf nor the bat-eared fox increases the variety of its diet with increasing drought. This suggests that there is sufficient invertebrate food for both in the central Namib Desert, even though the rainfall is low. Hence there is little food competition between the carnivore species which depend upon invertebrate food there (Bothma et al. 1984).

In terms of foraging in the central Namib Desert, the foraging ranges of the aardwolf and the bat-eared fox overlap, while the black-backed jackal forages over a larger area, only at times passing through the foraging range of the aardwolf and bat-eared fox. No data on foraging range are available for the Cape fox. Competitive interactions between these four carnivores while foraging are rare, however. Different times of onset of activity and emergence from the burrow also limit competition, even when the same foraging range is used. The aardwolf becomes active when its main food becomes active, as does the bat-eared fox, but this happens at different times of the day. The Cape fox is strictly nocturnal, as is its main prey. Much of the food-niche separation reflects the influence of the use of different micro-habitats on the diets of these carnivores. Overall, enough separation occurs in the diet, use of space and activity times to allow the sympatric co-existence of the aardwolf, black-backed jackal, bat-eared fox and the Cape fox in the same arid habitat in the central Namib Desert (Bothma et al. 1984; Nel and Loutit 1986).

Along the Skeleton Coast, competition for food plays an insignificant role in the feeding ecology of the black-backed jackal. Nevertheless, the kelp gull is possibly the most important competitor of the black-backed jackal for food there, as both feed on seals and cormorants. However, this competition is largely indirect as the gulls usually forage when the jackals are absent (Hiscocks and Perrin 1987). On Nxai Pan, bat-eared foxes and black-backed jackals apparently co-exist without competition (Koop and Velimirov 1982).

Elsewhere in its range, the bat-eared fox of the Serengeti does not appear to compete directly with other carnivores, although it may fall prey to them from time to time (Lamprecht 1979) as in the southern Kalahari (Bothma and Le Riche 1984, 1986, 1994a). In the Masai Mara National Reserve, bat-eared foxes ignore smaller carnivores such as mongooses, but they are wary of large predators. However, a group of bat-eared foxes will readily mob predators such as the spotted hyaena, black-backed jackal, slender mongoose *Galerella sanguinea* and white-tailed mongoose *Ichneumia albicauda* when they come too close to a den (Malcolm 1986). In the presence of such predators, bat-eared foxes "stot" in high, stiff-limbed leaps. This behaviour is unknown in other canids (Lamprecht 1979). The bat-eared fox is unique in its foraging methods, insectivorous diet, home range overlap and ability to live at high population densities. These allow it to live in arid areas and survive competition from other small, sympatric carnivores. They easily colonize new areas as soon as they become available (Malcolm 1986).

CHAPTER 3

Family Felidae

3.1
Taxonomy, Distribution, and Status

Many cat species occupy diverse habitats. The same species sometimes ranges from deserts to tropical forests. This requires great adaptability, a characteristic of all cats. All felids rely on their physical environment for successful prey-use, which is a major facet of survival (Kruuk 1986). They are also more strictly carnivorous than other carnivores, and markedly more so than canids (Kleiman and Eisenberg 1973).

Felids spread from Europe throughout Africa. The Eurasian dirk-toothed cat *Megantereon cultridens*, for example, is found in South African fossil deposits along with a form of leopard *Panthera pardus* which was the most common predator of its time, although it was larger than modern leopards. This was some 1.8 to 1 million years ago. The cheetah *Acinonyx jubatus* is also present in these fossil deposits (Turner 1993). Today the family Felidae is diverse and occurs over much of the world. The felids are divided into three subfamilies, the Acinonychinae which is represented by the cheetah only, the Felinae and the Pantherinae. Eighteen genera and 36 species are currently recognized (Wozencraft 1993).

The threat to the conservation of all 36 cat species excluding the Pampas cat *Oncifelis colocolo* is such that they are either listed in Appendix I or II of CITES (Wozencraft 1993). Furthermore, the following cat species or subspecies are considered by the IUCN to be critically endangered (Nowell and Jackson 1995): the Asiatic cheetah *Acinonyx jubatus venaticus*, Florida panther *Puma concolor coryi*, eastern cougar *Puma concolor cougar*, Anatolian leopard *Panthera pardus tulliana*, Arabian leopard *Panthera pardus nimr*, Amur leopard *Panthera pardus orientalis*, North African leopard *Panthera pardus panthera*, Amur tiger *Panthera tigris altaica*, Sumatran tiger *Panthera tigris sumatrae*, and the South China tiger *Panthera tigris amoyensis*. In addition, 11 more cat species or subspecies are considered endangered. They are: the northwest African cheetah *Acinonyx jubatus hecki*, North African serval *Lynchailurus serval constantinus*, Iberian lynx *Lynx pardinus*, iriomote cat *Prionailurus bengalensis iriomotensis*, Asiatic lion *Panthera leo persica*, Caucasus leopard *Panthera pardus ciscaucasia*, Sri Lankan leopard *Panthera pardus kotiya*, North Chinese leopard *Panthera pardus japonensis*, Javan leopard *Panthera pardus melas*, tiger *Panthera tigris* and the snow leopard *Uncia uncia* (Nowell and Jackson 1995). Further details on the status and distribution of the world's cats can be found in Nowell and Jackson (1996).

Ten cat species occur in Africa, but only the black-footed or small spotted cat *Felis nigripes* and the African golden cat *Profelis aurata* occur in Africa only. The leopard *Panthera pardus* has the widest range of all large cats, occurring from deserts to moist tropical forests in Africa, the Middle East and Far East, northwards to Siberia and southwards to Sri Lanka and Malaysia (Hes, 1991). There has recently been a suggestion that the current 27 subspecies of leopard be reduced to 8, but this suggestion is not yet generally accepted (Anonymous 1996a).

The cheetah belongs to a subfamily that was previously much more diverse, and even had two extinct species in America (Caro 1994). As recently as 2000 years ago, four cheetah species still roamed open habitats in Africa, Asia, Europe and North America (Hunter 1996). The extant cheetah has been the subject of some controversy regarding its poor conservation future because of its relative genetic poverty. Of the current cheetah subspecies, the Asiatic cheetah probably numbers less than 50 individuals in Iran, while of the northwest African cheetah there are only several dozen left in the mountains of southeastern Algeria and possibly southern Morocco, and some 150 on the Aïr and Termit Mountains in Niger (Nowell and Jackson 1995, 1996). The cheetahs of the Sahara Desert are smaller and paler than cheetahs elsewhere (Nowell and Jackson 1995). Genetically, the cheetah is interesting for the occurrence of the so-called King cheetah genetic coat colour variation which is mainly found in the more arid parts of southern Africa (see below).

It is now accepted that all cheetahs are genetically similar with virtually no genetic diversity (eg. Hunter 1996), having a mean polymorphism of 3.2% and a mean heterozygosity of 0.013 as opposed to 14.7% and 0.036 respectively in other felids (Caro and Laurenson 1994). Another study shows no allelic diversity in 55 cheetahs at 52 loci with a polymorphism of 0% and a heterozygosity of 0. This has to be compared with the considerable allelic diversity in seven other felids which have a polymorphism range of 8 to 28.8% and a heterozygosity range of 0.029 to 0.072. These values are typical of mammals generally (Hunter 1996).

The cheetah's skull characteristics also reveal a high level of asymmetry, with the left, lower jaw longer than the right one. Such asymmetry is a common result of inbreeding, and the cheetah is the most genetically impoverished species ever examined. This is probably the result of a series of population bottlenecks, one of which occurred about 6000 to 20 000 years ago. It may also be because of a great interchange between subpopulations (Caro 1994; Nowell and Jackson 1996) at a time when many other large mammal species became extinct (Hunter 1996). The factor(s) which would or could have led to these bottlenecks are unknown (Nowell and Jackson 1996). In comparison, the cheetah's closest relative, the puma or cougar *Puma concolor*, has abundant genetic variation (O'Brien 1994).

The cheetah's lack of genetic diversity seems to be linked to semen with spermatozoal concentrations seven to 10 times less than those of domestic cats, and abnormal spermatozoa levels of 70 to 80% compared to 29% in domestic cats. Such lack of genetic diversity may explain the 29.1% rate of infant mortality in captive bred cheetahs (Hunter 1996). However, the cheetah has survived for at least 10 000 years despite its genetic impoverishment. Merola (1994) therefore believes that genetics are no real threat to cheetah conservation. Caro and Laurenson (1994) add that it has yet to be shown that inbreeding has led to the extinction of any wild population of cheetahs. They also believe that genetics have been overemphasized and environmental causes underemphasized in cheetah population dynamics. The real threat to the cheetah therefore seems to lie in a combination of genetic and environmental factors. Moreover, successful cheetah breeding by means of artificial insemination and other ways (Lindburg 1989), contributes to the cheetah's survival.

The King cheetah coat pattern mainly occurs in arid southern Africa. It is one of several known mutant skin patterns in cheetahs, and it is controlled by a single recessive gene (Lindburg 1989). Once regarded as a separate species *Acinonyx rex*, this genetic variant of *Acinonyx jubatus* has softer, longer and silkier hair than normal. The spots are also replaced by dark bars or stripes on the dorsal surface (Van Aarde and Van Dyk 1986). Almost all King cheetah sightings are from relatively arid areas in southern Africa between latitudes 17 and 25 °S and longitudes 24 and 33 °E. This area includes Zimbabwe, Botswana and South Africa, an area where the cheetah's coat pattern varies much between localities. The King cheetah coat pattern was at one time erroneously thought to represent an aberrant leopard or even a leopard-cheetah hybrid (Hills and Smithers 1980). There are recent reports of a King cheetah patterned skin from Burkina Faso too (Frame 1992; Nowell and Jackson 1996).

Of the other cats, the Asiatic lion *Panthera leo persica* of western India also has low genetic variation, leaving it vulnerable to survival. Genetic studies suggest that these lions separated from African lions 55 000 to 200 000 years ago. Visible differences include the belly fold of Asiatic lions, which is rare in African lions, sparse manes in male Asiatic lions, and even differences in skull shape (Johnsingh and Chellam 1991). In this context it is interesting to note that the Somali lion, which is geographically the closest of the African lions to the Asiatic lion, lacks a mane even in large, adult males (Fagotto 1985). In contrast, the two extinct subspecies, the Barbary or Atlas lion *Panthera leo leo* of Morocco, and the Cape lion *Panthera leo melanochaitus* of southern South Africa, occurred on the northwestern and extreme southern opposite ends of the continent from the Somali lion. These subspecies had dense, well-developed manes extending over the shoulders and chest to the belly (Mazak 1964, 1970; Smithers 1983; Anonymous 1996b). The Asiatic lion population, although resident in a relatively humid area with a rainfall of 650 mm in the east and 1000 mm in the west, also went through a major genetic bottleneck in the late 19th and 20th centuries with an estimated population low of no more than 20 individuals compared with an estimated 221 individuals in 1990 (Chellam and Johnsingh 1993). By 1995 there were an estimated 304 Asiatic lions, all in the Gir Lion Sanctuary of India or its adjacent areas (Singh 1995).

3.2 Social Behaviour

3.2.1 Group Formation

Felids range from totally solitary to highly social animals. Most wild cats are solitary as adults for most of their lives, the African lion (Fig. 3.1) being an exception. In the Kruger National Park, the largest lion pride recorded by Smuts (1982) is 21, but Aiken (1987) has recorded as many as 39, but possibly more, lions in a pride there. One or two large males usually dominate the pride (Grobler et al. 1984). In the Etosha National Park, an area receiving <400 mm of rain per year, the mean density of lions is lower (1.6 to 2 subadult and adult lions per 100 km^2) than anywhere else in Africa, yet the mean pride size is 8.8 (range 5 to 16) (Stander 1991). This compares with the modal group size of 2, and a mean pride size of 2.8 lions (maximum 13) which is found in the more mesic Serengeti National Park (Schaller 1972). In the Etosha National Park, lion densities correlate directly with the number of available water points (Stander 1991). The lions there also live in social groups with no clear social hierarchy among pride lionesses as is found elsewhere in Africa (Stander 1992a). In the southern Kalahari, which is more arid than Etosha, the density of lions is estimated at 1.3 lions per 100 km^2 (Smuts 1982) and this density and the mean pride size of 4.2 individuals (maximum pride size 21) are smaller than in the lions of Etosha (Eloff 1973a, c). Density and group size apparently were also low in the Barbary lion (Mazak 1970). It is believed that these variations in group size reflect a selective capacity of social animals to adjust group size to a hierarchy of ecological conditions. Primarily, these differences reflect the temporal and spatial patterns of resource availability, which in turn influence foraging efficiency, mating probability and reproductive success (Caraco and Wolf 1975).

A further factor influencing optimal lion group size in the Savuti Marsh may be the influence of competition by other predators such as the spotted hyaena. There, the lion prides suffer from a shortage of adult males: one male per 5.7 females compared with one male per 1.7 to 2.4 females in the southern Kalahari and Serengeti. Consequently, groups of female and subadult lions lose almost 20% of their food to spotted hyaenas. Such losses are most frequent for lions living in smaller groups (Cooper 1991). Increased food availability in lions is known to lead to increased pride size because of higher cub survival rates. However, it also leads to new prides being formed (Hanby and Bygott 1979). Therefore it can be expected that lion group size in arid areas, with relatively sparse food resources, will not be overly large. Also, as in the Serengeti, the onset of the dry season may cause larger lion groups to break up into smaller groups which are energetically more efficient (Caraco and Wolf 1975).

The only other felid that occasionally forms social groups is the cheetah. There male offspring born together may remain together for life to form a male group. Also, a female and her dependent subadult offspring move around together. However, non-relatives may occasionally join forces if such a larger group allows the take-over of an existing territory or the fending off of a challenge from intruders (Caro 1994). Larger group size in male cheetahs seems to be related to greater prey availability as happens with lions and cheetahs in the Serengeti (Caro and Collins 1987a,b). Moreover, male sibling cheetah groups seldom number more than three to four individuals because there is seldom more than such a number of male siblings in a given litter (Caro and Collins 1987a,b). Male cheetahs in the arid southern Kalahari, however, seldom move around in groups of more than two individuals (Labuschagne 1979). Grouping in male cheetahs is most probably a response to intraspecific competition, with greater food intake in groups being a secondary consequence (Caro 1994). However, group hunting as such in felids will be discussed later.

3.2.2 Nomads

In the Kalahari Gemsbok National Park, many subadult male lions leave the park and therefore do not remain in established prides (Smuts 1982). These nomads wander widely, often following migrating game in the Kalahari as they do in the Serengeti (Schaller 1972). However, in the Etosha National Park, solitary hunting lions are the only ones who do not meet the estimated minimum daily food requirement per lion (Stander 1992b). In the Serengeti, any male lion cubs which survive to adulthood leave the pride together, and remain together as a group of breeding males in a different pride (Bertram 1979). There the nomads of the lion population form about 20% of the entire population. It occasionally includes one, or a few, nomadic females of any age, which may settle in an area and build up a pride by recruitment of young. As environmental conditions improve, however, the number of nomadic lions decreases because fewer subadults leave their prides and more subadult females form new prides. Mortality is usually high among nomads (Hanby and Bygott 1979). In the southern Kalahari, adult male lions often move between prides which contain females and young of both sexes. At least three different prides shared a common area with a single roving male on one occasion there (Eloff 1973a). In the Savuti, adult male lions in resident prides are rare (Cooper 1991). Some females also become nomads, although these in turn may form small groups (McBride 1990).

Single, male cheetahs are at a great disadvantage in fights when attacked by a group of males. However, there is a lack of dominance in a group of males, who probably share in mating opportunities (Caro 1994). Nomadic, male cheetahs generally live shorter lives, are in poorer physical condition, and weigh less than territorial ones (Caro and Collins 1987b).

3.2.3 Communication

Most mammals display an array of vocal and visual communication systems, but especially so in the more social species. In some carnivores, such as the mongooses, this has led to highly specific messages about the location of food and predators (e.g. Beynon and Rasa 1989) and to intergroup activities in hyaenas (Mills 1990).

As most cats live solitary lives, vocal and visual communication between them has developed to varying degrees. In cats, complex expressive movements are highly developed, despite most cats being solitary (Schaller 1972). As the body postures and vocal expressions within one species are likely to be relatively similar from area to area, this section will deal more with chemical communication, which, intuitively viewed, may be influenced by the aridity of the environment and its sparser resources. However, no such evidence could be found in the literature.

Communication among African lions has been dealt with extensively by Schaller (1972) in Serengeti lions. There it includes vocalizations, facial expressions, postures and scent. Long-range signals in these lions mainly involve scent-marking and roaring, the intensity or frequency of which varies with circumstances (Schaller 1972). Both of these are also more frequent when intrusion pressure by nomadic lions is the greatest (Hanby et al. 1995).

Among the glands used by cats in scent-marking, the highly-developed ones above the tail, the anal and praegenital glands, the circumoral and submandibular glands and the glands between the footpads are used often (Verberne and Leyhausen 1976). As little is known specifically about scent-marking by lions in arid areas, the information on Serengeti lions (Schaller 1972) will mainly be summarized here as it should probably also occur in the lions of more arid environments. Furthermore, there are also long, dry periods in the Serengeti ecosystem.

Any lion will leave its tracks, urine and faeces in the environment as it moves about. These may serve as signals to other lions passing by later. Lions spray-urinate a musky-smelling substance on vegetation. This substance is expelled from their anal glands and is mixed with the urine squirted. The resulting mixture is used in small quantities, and distributed widely (Verberne and Leyhausen 1976). The mixture is squirted upward and backward at an angle of 30 to 40° for a distance of up to 3 to 4 m. Urine-spraying is predominantly done by male lions. When spray-urinating, male lions prefer to do so against vegetation, often a tree trunk at least 1 m high.

Several felines scrape on the ground or wipe with their feet (Verberne and Leyhausen 1976). Such scrapes or wipes are used by male and female lions for chemical communication. When scraping, a lion rakes the ground from two to 30 times with the claws of the hind feet, undoubtedly leaving a concentration of scent in the process. African lions often spray-urinate while scraping (Schaller 1972; Verberne and Leyhausen 1976) or do so immediately afterward. Such urination usually consists of spurts which often wet the animal's hind legs. There is evidence that lions can distinguish the scents of individuals within the pride, and between prides (Hamilton 1976). However, lions have never been observed to defecate on scrapes as tigers often do. This suggests that scraping in lions is not as important in scent-marking as it may be in tigers. The frequency of scraping in lions varies with the habitat. Male lions in prides spray urine or scrape at a constant rate throughout the year. Spraying and scraping by one lion frequently stimulates others to do the same. Previously used sites may become permanent landmarks. Lions of all ages also sharpen the claws of their forepaws by raking tree trunks, scarring some trees repeatedly. Such trees also serve as chemical sign posts (Schaller 1972).

Apart from the contact between adults when mating, and between a female and her cub, there is little physical contact between leopards (Fig. 3.2), although adults may occasionally fight (Scott 1985). Scraping, tree-scratching and urination are all forms of communication found in leopards (Smith 1978; Bothma and Le Riche 1984). In the Serengeti, the faeces of a leopard is not used to mark range boundaries as in the Masai Mara National Reserve (Scott 1985). However, ground scraping and tree-clawing occurs occasionally in the Serengeti, as it does in the Matobo National Park (Smith 1978). In the latter area, one specific tree was clawed repeatedly by a specific male leopard. Tree-clawing in the Masai Mara National Reserve leaves an odour-tainted visual and olfactory mark establishing the leopard's whereabouts to itself and to conspecific individuals. It also stretches the muscles involved and keeps the animal's claws sharp by peeling off loose flakes (Ewer 1973; Scott 1985). Female leopards in the Masai Mara claw trees regularly, occasionally scrape the ground and mark the edges of their ranges by defecating. These females also spray-urinate on low bushes (Scott 1985).

Tree-clawing by leopards varies between regions. In the Tsavo National Park for instance, tree-clawing was only observed once in a detailed study on leopards. However, spray-urination, faeces and scrapes were all used in marking the area. Scrapes are less common among African leopards than in those of India (Hamilton 1976). In the southern Kalahari, scraping is infrequent among leopards as elsewhere in Africa; but chemical communication is used extensively. There both sexes claw trees; females doing so less often than males. Regularly used scratch posts are rare, possibly because of the large ranges used by these leopards and the relatively infrequent use of a given area of habitat by a certain leopard over a given period of time (Bothma and Le Riche 1984, 1995).

Both sexes of leopard spray-urinate against shrubs or trees, and scrapes are often associated with urination (Eisenberg and Lockhart 1972). In the southern Kalahari, males repeatedly scent-mark trees by urination prior to mating. However, females seldom scent-mark trees and shrubs. In males, this activity seems to be aimed more at females than at other males. Both sexes use trees as rubbing posts, males doing so more often than females. Rubbing is done presumably for grooming, and coarse-barked *Acacia erioloba* tree trunks are clearly selected as rubbing posts over other trees (Bothma and Le Riche 1995). In the Kruger National Park, a leopard sometimes ascends a tree and urine-marks the branches (Pienaar 1969).

Apart from scent, the leopard female uses her tail as a visual guide or signal when leading her cubs somewhere (Schaller 1972). Leopards also roll on the dung of other species, or do so on the ground, possibly where other animals have urinated. Whether this is a form of communication is unknown (Scott 1985; Bothma and Le Riche 1984).

Tree-scratching has never been observed in cheetahs in the arid South African bushveld (Kruger 1988), and rarely so in East Africa (Eaton 1970a). However, cheetahs in the southern Kalahari urine-mark trees on average every 1.6 km moved, using the same tree repeatedly to do so (Labuschagne 1979).

Verberne and De Boer (1976) refer to flehmen in a male cheetah after smelling a female's urine, suggesting some sort of chemical communication. Male cheetahs are as capable of directional urination as tigers and lions. Cheetahs pass small quantities of urine throughout the day onto objects in their range. Female cheetahs usually spray-urinate widely and frequently, but seldom urinate heavily. The urine of a female cheetah in oestrus relays a definite message to males (Eaton 1970a). Marking locations of cheetahs are regularly marked over and over in the same location, but the density of such markings increases with the openness of the environment (Eaton 1970a). Cheetahs in the southern Kalahari regularly defecate next to specific trees in a scent-marking action (Labuschagne 1979). Cape wild cat males also spray-urinate against bushes (Mills et al. 1984).

3.3 Reproduction

None of the cats maintain permanent pair bonds (Eltringham 1979). The urine and cheek gland secretions of female domestic cats generally appear to contain pheromones which inform males about their hormonal phase. The same may be true for cats in the wild. Males commonly show flehmen in response to these messages. Both male and female cats spend much time in exploring the urine and skin gland secretions of foreign cats who may come into their range (Verberne and De Boer 1976).

3.3.1 Lions

Nomadic, male African lions often move between prides consisting mostly of females and their offspring. In the southern Kalahari there is no clear mating season, but lions stay together for several days, copulating two to three times every hour. Occasionally, two males may share a female, but this is rare. Whilst mating, lions seldom hunt (Eloff 1973a).

Infanticide by immigrating males occurs regularly in African lions. Males who take over a pride, kill the resident infants when they first encounter them, but are unlikely to kill their own offspring subsequently. Such infanticide is clearly disadvantageous to the females. These females, such as those in the Serengeti, adapt in various ways to reduce their cub losses due to infanticide. Such adaptations involve various behavioural responses aimed at minimizing losses if they cannot be prevented at all. Females whose young are killed through infanticide, resume sexual activity soon after the killing of the cubs (Packer and Pusey 1983). This often leads to synchronised breeding of all females in a pride which has been taken over by a new male (Packer and Pusey 1987).

The act of infanticide by immigrating lion males could be a reproductive strategy allowing such males to stop females from investing in the offspring of evicted males. Females, often assisted by other females, respond to this threat aggressively, attacking the male lion. In doing so, they allow some of the cubs to escape. Sometimes older cubs one-and-a-half to two years of age are evicted from the pride without being killed. They may be accompanied by some of the females, resulting in only a partial take-over of a pride. However, despite these adaptations, cub losses to infanticide may be high (Packer and Pusey 1983).

The gestation period of an African lion is about 90 to 110 days (Schaller 1972; Rudnai 1973). Large foraging groups of six to eight female lions in the Serengeti ecosystem enhance reproductive success. This happens because the births tend to coincide, and the females protect and feed each other's young, leading to increased cub survival (Caraco and Wolf 1975) Nevertheless, only 20% of the matings there lead to successful conception (Schaller 1972). In the Serengeti and the Ngorongoro Crater, the sex ratio of cubs born during the first 300 days after a male take-over of a pride favours males. It also differs significantly from that of cubs born subsequently. Therefore, more males are born when females breed synchronously after a male take-over (Packer and Pusey 1987). A lion cub is weaned when 10 weeks old (Grobler et al. 1984).

All pride males in the semi-arid bushveld of the Kruger National Park are between five and nine years old. No males younger than five years old associate sexually with the females, despite the fact that males become sexually mature at a median age of 30 months. Females become sexually mature at 48 months of age, and by the age of 5.5 years, all females sampled there had ovulated previously (Smuts et al. 1978).

Of all parous female lions in the Kruger National Park, 20% are pregnant at any given time, 21.7% are lactating and 6.7% are sterile. In one study, the two oldest females capable of reproduction were 12 and 13 years old. Lactation continues for 8 to 9 months post-partum and the mean litter size is 3.08 at birth, 3.02 when 1 year old and 2.71 when between 1 and 2 years old (Smuts et al. 1978). These parameters are similar to the mean litter size of 3 at birth in the southern Kalahari lions (Eloff 1973a), and to those in the Nairobi National Park (Rudnai 1973). However, they are higher than that of 2.3 (maximum: seven) found in the Serengeti where litter size of lions seems to be independent of food abundance (Schaller 1972). The overall conclusion that litter size in African lions is independent of food abundance agrees with the presence of large litters as are found in some arid areas.

Pregnancy control in the lions of the Etosha National Park was practised successfully by implanting a tiny capsule into the neck muscle of a lioness. This capsule releases a constant, minute amount of pregnancy-simulating hormones. This prevents ovulation for up to 5 years (Berry 1983). Under given circumstances, lions are also able to increase their reproductive rate. Cub survival in lions seems to be related to the abundance of food resources (Schaller 1972; Rudnai 1973; Smuts et al. 1978). The presence of abundant food decreases the time that mothers have to spend away from their cubs while hunting. Consequently, the culling of lions in the Kruger National Park, which significantly decreased lion numbers and led to more food available per given remaining lion, also led to increased reproduction (Smuts 1978b).

Male lions play no direct role in cub rearing in the Savuti Marsh (McBride 1990). A lioness never brings food to her cubs (Scott 1985) and these eat more relative to their weight than adult lions do (Smuts 1979). Lion cubs in the Savuti do eat meat from an age of 2 weeks, however, if available and while still suckling (McBride 1990). However, lion cubs usually only start to hunt when close to a year old (Schaller 1972; McBride 1990). Subadult males leave their natal pride at between 27 and 36 months of age in the Kruger National Park, but females either remain with the natal pride, or emigrate as early as 24 months of age (Smuts et al. 1978). A lion reaches its asymptotic weight at 7 to 8 years of age (Van Jaarsveld et al. 1988). The mortality rate of lion cubs in the southern Kalahari is high, probably the highest in Africa, particularly as the prey base there is relatively low (Eloff 1973a, 1980). Hofer and East (1993a) believe that, apart from spotted and brown hyaenas, the cubs of other carnivores cannot survive without nutrition for more than 24 to 36 h. However, it often happens that a lioness in the southern Kalahari will leave her cubs unattended for 2 to 3 days while out hunting. During this time the cubs are at the mercy of the harsh environment as well as marauding predators. Despite this, they can and do survive, as can leopard cubs there (Eloff 1973a). Although it is known that allo-suckling occurs in lions, other lactating females must still be available in a pride if the cubs are to survive in this way. This is especially so if their own mother should die or abandon them (Eloff 1980). Nevertheless, in the southern Kalahari, predators and starvation are the main causes of mortality in lion cubs (Eloff 1973a).

On the arid Serengeti Plains west of the Ngorongoro Crater, the low survival rate of lion cubs is also related to the harsh environment. It is less than half of what it is in the wetter and adjacent Ngorongoro Crater. Most cub mortality on these arid plains occurs before the cubs are a year old (Hanby et al. 1995). An overall cub mortality of 67% occurs for lions in the Serengeti (Schaller 1972). In the central part of the Kruger National Park, lion cub mortality is estimated at 29%. Only 40% of the cubs in the Etosha National Park reach maturity (Pienaar 1960; Eloff 1980). In the southern Kalahari, lion cubs may also die of scorpion stings, particularly as the area is rife with them. The disease rickets is another cause of lion cub mortality in this harsh environment (Eloff 1980).

Further evidence for the influence of prey abundance on lion cub survival is found in the Serengeti. When conditions which prevailed from 1966 to 1969 are compared with those from 1975 to 1977, the available prey biomass doubled. At the same time the number of cubs in each lion pride increased significantly. This was probably because the cubs survived better, and the fact that that more cubs remained with their natal prides as there was enough food. Cub survival to two years of age increased from 14% to 44% in this interval. Once a lion cub has reached 2 years, its chances of premature death decrease considerably (Hanby and Bygott 1979). It is possible that cub mortality may be a good indicator of population condition and the physiological condition of lions as in the Etosha National Park (Melton et al. 1987).

3.3.2 Leopards

There are no in depth data on the reproduction of leopards in the wild, and especially so in arid areas. The gestation period is 90 to 106 days (Grobler et al. 1984; Scott 1985; Le Roux and Skinner 1989; Hes 1991). Oestrus lasts six to seven days, and will recur every 20 to 50 days. A female in oestrus travels widely and calls often. Hormonal changes are relayed to males by way of substances in her urine (Scott 1985; Bothma and Le Riche 1995). There is no clear mating season and the mating pair stays together for only 24 to 48 h (Scott 1985; Bothma and Le Riche 1989). When mating, leopards copulate frequently, and a mean interval between copulation of 30 min has been recorded (Baker 1982). Copulation itself is brief (Scott 1985). Female leopards produce their first young when about 3 years old. In the arid bushveld of South Africa, there is an interval of 17.1 months between successive litters (Le Roux and Skinner 1989).

The cubs of leopards in the southern Kalahari are usually born in aardvark burrows taken over by the leopard. The cubs spend their early life under dense vegetation such as the shepherd's tree *Boscia albitrunca*, or in the burrows of mammals like the springhare and various mongooses (Bothma and Le Riche 1986). In East Africa, leopard cubs are known to home in on their dens by its odour. The eyes of a leopard cub are closed at birth, opening at 6 to 10 days of life (Grobler et al. 1984; Scott 1985). The female stays close to the cubs for the first few days, then moves about to hunt (Bothma and Le Riche 1986).

Cubs start to wean when about 42 days old, start to feed on meat when about 4 months old, being led to food by the female (Scott 1985), and will begin to kill their own prey when about 5 months old (Smithers 1983). They will eat meat brought to them by their mother before reaching 8 weeks (Scott 1985), but remain totally dependent on their mother for the first 11 to 18 months of life (Scott 1985; Le Roux and Skinner 1989). After this time the cubs continue to associate with their mother on a less permanent basis (Scott 1985). Litter size usually varies from one to three (Bothma and le Riche 1986; Hes 1991) but it can be as high as six (Hes 1991), although such big litters are rare.

Unweaned cubs are moved frequently to new sites of seclusion (Scott 1985; Le Roux and Skinner 1989). In the southern Kalahari, female leopards only cover their faeces and kills with sand when close to their own young cubs, possibly in cub-defence. Other defecations are not covered (Bothma and Le Riche 1994a).

Female leopards make a chuffling sound consisting of three or more short, sharp puffs as a vocal greeting to communicate with their cubs. Cubs remain relatively inactive when the female is away, but other predators such as eagles, lions, spotted hyaenas, jackals, baboons and even male leopards still regularly find and kill them (Scott 1985). In more prey-abundant areas, females usually return to their cubs every day, even when hunting is unsuccessful (Scott 1985). Nevertheless, even in the prey-rich Chitwan National Park in Nepal, a female leopard may be absent from her cubs for a day or more, her period of absence being determined by the time required to search for and kill prey (Seidensticker 1977). In the prey-poor southern Kalahari, females may leave their cubs alone for as long as 2 nights while out hunting, considerably increasing the chances of cub mortality (Bothma and Le Riche 1984, 1986). When a female is close to her cubs, she will suckle them several times per day (Scott 1985). Nevertheless, it is clear that among leopard cubs there is an adaptation for a "feast and famine" food regime which operates from a very early age (Seidensticker 1977).

Leopard cubs at play use a sequence of instinctive behaviour patterns common to all cats. However, from an age of about 8 months, the cubs start to socialize less with each other, spending little time together after an age of 11 months or more, being fully grown and solitary by 30 months of age (Scott 1985), and sexually mature at a mean age of 3 years (Hes 1991).

Mortality in leopard cubs usually is high (Scott 1985) and few cubs survive to adulthood. Young, male leopards leave their natal range when 15 to 16 months old, dispersing gradually until they can establish an own territory (Sunquist 1983). In the southern Kalahari, a subadult male leopard dispersed over a straight line distance of 112.6 km over a period of three months before establishing a permanent home range (Bothma unpubl.). The timing of dispersal in young leopards is a function of a variety of social and environmental factors because the age at which leopards disperse varies from area to area (Sunquist 1983).

3.3.3 Cheetahs

Cheetahs have considerable problems in reproducing successfully, including congenital mortalities (O'Brien 1994). In addition, they have extremely high cub mortalities (Eaton 1970a; Schaller 1972; O'Brien 1994; Caro 1994; Laurenson 1995). Cheetahs are induced ovulators with no marked peak in breeding. A high percentage of the spermatozoa show abnormalities, but the effect of these on fertility is still largely unknown (Caro 1994). Cheetah females are highly selective of their mates. Gestation lasts for 90 to 95 days, and the cubs are born with closed eyes (Grobler et al. 1984).

In the wild, one to six cheetah cubs are born to a litter, and they remain dependent on their mother for more than a year, although they are weaned when two to three months old (Mills 1990; Caro 1994). Most adult females give birth, but even in a prey-rich area such as the Serengeti only 28.8% of all cubs born eventually survive to emerge from the den, and only 8.8% survive till four months old. Predation by lions is the primary mortality agent. Of all cubs born in the Serengeti, 4.8% have a chance to reach independence (Caro 1994).

About 50% of all cheetah cubs in the southern Kalahari die before they are 6 months old, primarily because of predators and starvation. The known predators of cheetah cubs in the southern Kalahari include the black-backed jackal and occasionally the brown hyaena *Parahyaena brunnea*. These predators usually find and kill cheetah cubs while the female is away hunting. As a given female cheetah once left her cubs alone and unattended eight times in 22 days to go away and hunt in the southern Kalahari, cheetah cubs there are highly vulnerable to predation (Labuschagne 1979). Of all cheetah cubs born in the Serengeti, 7.7% are eventually abandoned by their mothers and starve to death. The difficulty of successful hunting and the availability of prey at certain times of the year are major reasons for such abandonment. Where other predators are virtually absent, such as on farmland in Namibia, however, cheetahs do well and litter size at independence can be as high as four.

This clearly illustrates the devastating potential effect of predation on cheetah cub survival in a variety of ecological conditions (Laurenson 1995).

Cheetah cubs start moving with the female when six weeks old, reach sexual maturity when 15 months old and leave their mother when 18 months old (Grobler et al. 1984). Young cheetahs learning to hunt already possess a fair degree of innate predatory behaviour, but the training which they receive from their mother is essential for the eventual development of the entire predatory sequence (Eaton 1970b). Hunting success in cheetahs increases with age (Caro 1994).

Female cheetahs rapidly conceive again following the loss of unweaned cubs. Conceptions when food is abundant are usually most successful. Cubs losing their mothers sometimes join unwilling females and steal food from them to survive. Cheetah cubs remain with their mothers until about 18 months old. They are nursed for the first 4 months, a period in which their mothers double their food intake in the Serengeti (Caro 1994).

On the Serengeti Plains, cheetah cub recruitment is not enough at present to maintain the cheetah population there (Laurenson 1995). This suggests a variable pattern of recruitment and survival over considerable time for the population to have survived to date.

3.3.4 Other Cats

The African caracal occurs extensively in arid areas, but little is known about it there. However, it seems that the sexes only associate when mating (Pringle and Pringle 1979). The litter size varies from one to four (Grobler et al. 1984). In the arid Eastern Cape Province areas of South Africa, young are born in late autumn, spring and summer (Pringle and Pringle 1979). The gestation period is 76 days (Grobler et al. 1984). In the even more arid West Coast National Park, caracal litters are born mainly in summer, with a mean litter size of 2.25 (Avenant 1993). The eyes are closed at birth. Kittens stay with their mothers for their first 4 months of life (Avenant 1993). In the arid Karoo regions of South Africa, caracal in the Mountain Zebra National Park disperse form their natal ranges when 7 to 9 months old. One young, dispersing male there moved 10 km in 8 months, and another one 22 km in two months before settling in an own territory (Moolman 1986). In the Cape wild cat, litters of two to five are born in the wet season after a gestation period of some 58 days (Smithers 1971; Grobler et al. 1984). The kittens hide in thickets by day and only the female raises the young. For the black-footed cat, litters of one to three are born in burrows, mostly in the summer. The kittens are cared for by the female only (Grobler et al. 1984).

3.4
Area Use, Movements, and Activity

The basic pattern of felid social organization is one in which males occupy large, exclusive ranges that encompass several females' ranges. In range use, the females in turn are keyed to resource availability (Sunquist and Sunquist 1989).

3.4.1
Lions

Contrary to popular belief, the African lion (Fig. 3.1) typically is an animal of dense, dry habitat (Smuts 1982). African lion individuals can either be nomads or residents and members of a pride living in a given area. These categories are not mutually exclusive, and a nomad can become a resident and vice versa, but the two ways of life differ vastly. A pride occupies a given range or territory, and a nomad one of up to ten times that of a pride. The size of the range of a lion pride in the Serengeti ecosystem is related to the size of the pride and the resources available: pride territories often overlap extensively in space but usually not in time (Schaller 1972).

Nomadic African lions in the Serengeti wander widely, often following the migrating game. Nomads there defend no given area and are remarkably tolerant of strange lions (Schaller 1972). In the southern Kalahari, most lions tend to have territories, but territory size varies. One pride there hunted in a known area of 120 km^2, another in one of 273 km^2. However, there are also

Fig. 3.1. The African lion is a typical animal of dense, dry habitats

nomadic lions in the southern Kalahari which move between prides or may stay alone for some time (Eloff 1973a). In the Etosha National Park, lion prides use areas ranging from 150 to 2075 km^2 in size. Area size used there is related to pride size, the larger the pride, the larger the area used (Stander 1991). In the Savuti Marsh, lions occupy territories as large as 300 km^2 for life, but the territories may shift over time and the size of each territory depends upon the number of prey present on a year-round basis (McBride 1990).

Some Asiatic male lions may live in coalitions of two to six animals. These actively defend a territory which usually includes one or more prides of females and young (Johnsingh and Chellam 1991). In the Gir forest of India the mean range of prides (50 km^2) is much smaller than those of lions in the southern Kalahari (Eloff 1973a). However, they are larger than the 20 to 30 km^2 territories of prides in the Serengeti (Hanby and Bygott 1979).

Lions in the southern Kalahari at times hunt in definite circuits, one pride completing a circuit of 120 km in 4 days. The greatest individual daily distance moved by any lion in the southern Kalahari, is a female who moved 41.2 km in one night. A pride of three females moved a mean of 22.3 km per night on eight consecutive nights, the greatest mean distance recorded there (Eloff 1984). This compares with up to 21.5 km in 24 h in the Serengeti (Schaller 1972). Although covering great distances when moving, lions seldom hurry, and often lie or sit down to rest for a while (Eloff 1973a). In the Etosha National Park, the mean daily distance moved by prides of lion is 13.2 km in 24 h (Stander 1992b), a distance which is similar to that of the lions in the southern Kalahari.

The African lion seldom climbs trees, but it may do so occasionally to escape danger. However, in the Lake Manyara National Park, lions often rest by day in trees, possibly to evade biting flies. The southern Kalahari lions also occasionally rest in trees, possibly to avoid ground-living, biting invertebrates such as the sand tampan *Ornithodoros savignyi* (Smuts 1982).

Female lion pride members in the Serengeti defend their territories against female intruders, and males against males (Hanby and Bygott 1979). In the Kruger National Park, the territory tenure of a male lion starts at an age of 5 years, and continues at most for a few years (Smuts et al. 1978).

Lion numbers, and hence their density, are related to resource abundance and availability. This means that they also have to be influenced by aridity. In the resource-rich Serengeti ecosystem, there is one lion per 6.3 km^2 in the woodland areas (Schaller 1972), compared with one lion per 12.5 km^2 in the Selous Game Reserve (Rodgers 1974), one lion per 7.7 km^2 in Kruger National Park (Smuts 1982), one lion per 35.7 km^2 in the Etosha National Park (Stander 1992a) and one lion per 76.9 km^2 for the southern Kalahari. The arid Masai Steppe of Tanzania has the lowest known lion density of one lion per 306 km^2 (Mills et al. 1978). In the Etosha National Park, there is a clear and direct correlation between lion pride size, and hence density, and the number of waterpoints, and hence prey availability (Stander 1991).

3.4.2
Leopards

The leopard (Fig. 3.2) is an agile cat which moves with ease and can leap gaps of up to 6.0 m (Scott 1985). It has adapted to life in diverse environments (Smithers 1983). This habit also reflects its available prey-base and hence its area use and movements. The leopard has an exclusive range for a given sex, which is probably a territory, although it is not as much defended against as avoided by other leopards (Eltringham 1979). Hes (1991) believes that these ranges are territories. Female leopards usually have overlapping ranges, shared with one male (Eisenberg 1970; Schaller 1972). Male leopards will become involved in fierce physical territorial disputes (Grobler et al. 1984). In moist, tropical areas such as Sri Lanka, leopards have small ranges, varying from 8 to 10 km² in prime habitat, and 29 km² in more marginal areas (Muckenhirn and Eisenberg 1973). This compares with ranges of 40 and 60 km² (maximum 262 km²) in two females in the Serengeti (Schaller 1972), 10 to 60 km² in the Tsavo National Park (Hamilton 1976), a mean of 18 km² (range 10 to 19 km²) in the Matobo National Park (Smith 1978), a range of 23 km² (possibly 33 km²) for a single female in the arid bushveld of the Sabi-Sand Game Reserve (Le Roux and Skinner 1989), and mean ranges of 2182.4 km² (n = 3; males) and 488.7 km² (n = 5; females) in the extremely arid southern Kalahari (Bothma et al. In Press).

Where prey is scarce, a leopard patrols its range regularly, unconsciously giving the prey a time of respite from predation pressure (Turnbull-Kemp

Fig. 3.2. The leopard has the widest range of all large cats

1967; Bothma and Le Riche 1984). Apart from food abundance and availability, the density of other competing predators influences the size of a leopard's range (Schaller 1972). In the Tsavo National Park, male leopards are at times as close as 0.8 km to each other, but are usually much further apart (Hamilton 1976).

Leopard ranges have a high degree of stability (Hamilton 1976). Some areas of the range, the core areas, are used more intensively than others (Smith 1978). In the southern Kalahari, the mean core-area size of adult female leopards is 26% of the total range, and 21% in males (Bothma, unpubl.). Transient leopards may occasionally move through a given leopard's range (Ewer 1973; Smith 1978). Individual leopards spend several months in the core area of their range, but at other times they are active elsewhere.

In the Sabi-Sand Game Reserve, a male leopard visits a given part of its range once every 17.8 days, and a female once every 8.8 days. However, this varies seasonally (Kruger 1988). In the southern Kalahari, leopards move significantly longer distances per 24-h period as hunger increases (Bothma and Le Riche 1990). The long-term, mean distances moved by leopards there are 13.4 km in 24 h for adult females and 14.2 km for males (Bothma and Le Riche 1984). On the basis of hunger, those leopards in the southern Kalahari, that had recently killed prey moved a mean distance of 10.1 km in 24 h; those that had gone without food for one day 14.2 km; those that had been without food for two consecutive days 20 km, and those that had had no food for three or more consecutive days, 21.8 km (Bothma and Le Riche 1990). The maximum known distance covered in 24 h by an adult male leopard in the southern Kalahari is 33 km and 27.3 km by a female with cubs (Bothma and Le Riche 1984).

Male leopards in the southern Kalahari, where summertime day temperatures reach the mid-40s in °C, are influenced significantly by ambient temperatures in their movements. On colder nights with minimum air temperatures of 0 °C or less, leopards of equal hunger status move a mean distance of 29 km in 24 h, compared with 19.1 km on nights of between 1 and 15 °C and 15.2 km on nights of 16 °C or more. Overall, male leopards with equal hunger status in the southern Kalahari move significantly longer distances (mean: 24.9 km) in 24 h in winter than in summer (mean: 17.2 km). These data probably reflect a combined influence of higher temperature and shorter nighttime in summer (Bothma and Le Riche 1994b).

When disturbed, leopard flight distances in the southern Kalahari are minimal, the mean flight distance being 91 m (range 15 m to 2.7 km) (Bothma and Le Riche 1993). Moving leopards in the southern Kalahari rest frequently, but they especially do so just after the onset and before the cessation of activity for any given day. The main daytime resting place is the shady and dense cover of the shepherd's tree, or an aardvark *Orycteropus afer* or porcupine *Hystrix africaeaustralis* burrow (Bothma and Le Riche 1984).

3.4.3
Cheetahs

Little is known about cheetah range use in arid areas. However, cheetah females usually have undefended ranges with considerable overlap, while territorial males are limited to small territories. In the Serengeti ecosystem, the females follow migratory game and the males remain behind. In consequence, there is a large mean range of 833 km^2 for female cheetahs there, compared with a small mean range of 37.4 km^2 for males. There are also, however, non-territorial males which range over areas with a mean size of 77.2 km^2. Territorial males occasionally make excursions outside their territories (Caro 1994). Some males also leave their territories during the dry season in search of food. There are large, vacant areas between established territories, even when all are occupied (Caro and Collins 1987a). Furthermore, the length of tenure of a territory increases with the group size of cheetahs involved. The territory occupants are usually all large animals for their age group (Caro and Collins 1987b).

Among the Serengeti cheetahs, body size and age are two crucial factors influencing whether cheetahs become territorial or not. Single males have to form groups with other males before being able to displace resident territorial males. Larger coalitions are, however, never displaced by smaller ones. Only adult males hold territories, but territorial males are not more likely to encounter females than non-territorial ones (Caro and Collins 1987b).

In the southern Kalahari, data on cheetah area use reveal a mean range in two adult males of 300 km^2, which is considerably larger than that found in East Africa. These two adult males moved a mean distance of 12.6 km in 24 h, often being inactive by day, when they usually rested in deep shade (Labuschagne 1979). An inactive period by day for cheetahs is also found in the arid bushveld areas in South Africa (Kruger 1988).

Cheetahs generally live in low densities compared with most other carnivores, but the reasons for this are still unknown. One possibility, however, is the high cub and juvenile mortality rates of cheetahs in the wild. These are mainly due to predation by other large carnivores (Eaton 1970a; Laurenson 1995). The Nairobi National Park has the highest density of one cheetah per 5.12 km^2 known anywhere (Eaton 1970a). The cheetah population in arid Namibia is considered to be the most viable one of any in Africa, possibly in part due to the relative absence of other large predators on farmland where most of Namibia's cheetahs occur and thrive (Phillips 1993).

3.4.4
Other Cats

Area-use by the caracal in arid regions has been studied in detail in two locations in South Africa. These are the Mountain Zebra and West Coast National Parks (Moolman 1986; Avenant 1993). In the mountainous Mountain Zebra

National Park, four adult caracal males used a mean range of 15.2 km^2 and four adult females one of 5.5 km^2, a significant difference. On adjacent sheep farms, three adult male caracals used a mean range of 19.1 km^2. This is significantly larger than that of males in the national park. The ranges of males and females overlap within and between the sexes. Some ranges are confined to the national park, others to the farming area, while yet others include both farmland and land in the national park in their range. Although some areas of each range are used more extensively than other parts, no data on core-area size are given. When a vacant range develops, a young caracal will establish itself there and the boundaries of that range and of the neighbouring ones may be adjusted (Moolman 1986).

In the arid West Coast National Park, with its succulent to semi-succulent scrub vegetation, the mean range of two adult male caracals was 27 km^2 and that of three adult females 7.4 km^2 (Avenant 1993). The West Coast National Park is more arid than the Mountain Zebra National Park and also does not have as high a herbivore biomass. Hence the prey base is poorer. The caracal of the West Coast National Park have larger ranges than in the Mountain Zebra National Park. In both cases, however, the range of adult males is approximately three to four times that of the females. In the West Coast National Park, caracal ranges also overlap within and between the sexes (Avenant 1993).

The mean spacing between resting adult male caracals in the Mountain Zebra National Park is 4 km, and 2.2 km in females. This difference is significant. The mean, straight-line distance moved by adult males in a given 24-h period there is 2 km. This is significantly different from the mean of 0.9 km moved by adult females. The males travel mainly along the brush of the upper slopes, while the females with cubs prefer the lower, dense riverine thickets where there is an abundant rodent prey base. These rodents are their primary food. Females without cubs use the same habitat as males (Moolman 1986).

The caracal of the Mountain Zebra National Park are often blamed for stock depredation on the adjacent farmland. However, their area use and density in the park clearly show that the park can never sustain or produce the number of caracal killed annually on the adjacent farms. Consequently, the farmers must be producing their own problem caracals which live on these farms, and the park is not a main reservoir for such caracal (Moolman 1986).

In a semi-arid area of Spain, the activity cycles of the Iberian lynx *Lynx pardinus* were studied in the Doñana National Park. Photoperiod there has a major influence on the circadian patterns of activity of the lynx throughout the year. In winter, maximum temperature, high barometric pressure (decreased activity), evaporation and moon phase all influence lynx activity significantly. In summer, maximum temperature (decreased activity), high barometric pressure, moon phase and hours of moonlight are important determinants of lynx activity. In spring, rainfall and maximum and minimum temperatures play a major role in lynx activity. Some climatological factors affect the lynxes directly in terms of activity, while others such as moonlight, act indirectly by influencing the activity cycles of the Mediterranean rabbit *Oryctolagus cuniculus* which is the lynx's main prey there (Beltràn and Delibes 1994).

Overall, the activity patterns of the lynx in the Doñana National Park is part of the adaptation of this felid to its environment. Lynxes there are more active as the photoperiod decreases. This results in seasonal activity cycles. However, the greatest photoperiod influence occurs in winter. Barometric pressure and evaporation changes also influence the circadian activity cycle of lynxes, and movements increase on warm nights in winter (Beltràn and Delibes 1994).

The nocturnal African wild cat *Felis silvestris* in an arid area near Elmenteita in Kenya has a mean range of 1.6 km^2 in a mixed habitat of dry woodland and grassland (Fuller et al. 1988). Like the African wild cat, the black-footed cat is also nocturnal (Smithers 1971).

3.5 Feeding Ecology

The evolutionary fitness of any predator depends largely upon the quantity and quality of its diet (Sunquist and Sunquist 1989). Felids are more strictly carnivorous than other carnivores, but especially more so than canids. This leads to more peaceful feeding in the canids in comparison with felids, where it is rare (Sect. 2.5.4; Kleiman and Eisenberg 1973).

Most felids are solitary. In terms of feeding ecology, a solitary existence is compatible with felid specializations. Such specializations have developed to deal with the ecological constraints which prey place upon being preyed on. These deal with aspects such as prey abundance, the temporal and spatial distribution of the prey, the size of prey, and prey defence. These factors all contribute to the most beneficial balance of energy expended while hunting, and energy gained from successful hunts (Sunquist and Sunquist 1989).

Felids use prey of a modal size, which varies considerably between different geographic areas, even for the same species. The modal prey size used in a given area is primarily related to prey availability and vulnerability. Usually, however, the modal prey size is less than the given predator's body mass (Sunquist and Sunquist 1989).

Like carnivores generally, felids are primarily predators, but some of them scavenge food whenever it is available. Most of the food of large felids comes from ungulates (Houston 1979). Felids eat little bone, and there seems to be a negative relationship between the size of a carnivore and the need to chew its food. Especially in socially feeding carnivores, such as lions, speed of ingestion may be more important than precision in the use of teeth. This lack of precision also explains the tendency towards breaking the teeth, particularly the canines, when feeding (Van Valkenburgh 1996).

In the big cats, the moods of prey-catching and consuming food seem to be relatively independent of each other (Leyhausen 1979). However, in the leopard, hunger is a motivation for hunting (Bothma and Le Riche 1990). Nevertheless, surplus killing of prey independent of hunger or nutritional needs is known in felids. Surplus killing is defined in Section 2.5.3. It usually is a waste of a carnivore's energy unless it stores its prey to return to feed later (Kruuk 1972a).

All felids rely extensively on the physical features of their environment when hunting. Almost any type of cover may be used to get as close as possible to a potential prey before making a final attack (Eltringham 1979; Leyhausen 1979; Sunquist and Sunquist 1989). Felids never rush at their intended prey with as much unrestrained eagerness as do many other predators (Leyhausen 1979). Identification of its prey is important if the predator is not to "bite off more than it can chew". Of the cats, the two stalking specialists in Africa are the lion and the leopard, both of which take prey with more stamina than themselves (Eltringham 1979). Intraspecific competition for kills among solitary felids seems to be rare. The possibility of felids meeting at a kill is reduced by temporal and spatial dispersal (Sunquist and Sunquist 1989).

3.5.1 Food

Lions. The African lion hunts opportunistically, but it does scavenge part of its diet, usually stealing carcasses from other predators (Houston 1979). However, predation by lions is complex, and the feeding habits of lions are conspicuously influenced by the size of the prey and the availability and density of food for scavenging (Kruuk and Turner 1967; Schaller 1972; Eloff 1973a,c, 1984). In the Serengeti ecosystem, lions on the plains have difficulty in acquiring sufficient food in the dry season (Hanby et al. 1995). In the whole area there, the size of the prey of lions appears to increase with the size of the hunting group (Kruuk and Turner 1967).

The African lion occasionally kills young elephants *Loxodonta africana* (Pienaar 1969; McBride 1990). Other carnivores are not usually used as food by lions as they are by leopards, even when the lions are hungry. However, carnivores such as spotted hyaenas, brown hyaenas, jackals, cheetahs and leopards may occasionally be killed by lions, but are usually not eaten (Schaller 1972; Eloff 1973a, 1984). Cannibalism occurs in lions, but the reason for this is unknown (Schaller 1972). In the Kruger National Park, lions remove over 50% of the biomass of prey killed by all the large predators (Mills and Biggs 1993). Lions in the Kafue National Park take a higher proportion of prey that is in poor condition, but fewer juvenile prey than any of the other predators (Mitchell et al. 1965).

In the Serengeti, prey selection by lions varies between seasons, across habitats and from year to year. Most of this variation can be attributed to the annual migration of some important prey, a migration which is driven by varying amounts and patterns of rainfall from year to year (Scheel and Packer 1995). In the southern Kalahari, moisture-rich plants such as the tsama melon *Citrullus lanatus* are eaten by lions to obtain moisture, especially when prey is scarce (Eloff 1973a). In the Savuti Marsh, lions eat prey from as small as a springhare to the occasional young elephant (McBride 1990).

Lions in the Namib Desert feed extensively on fur seals, but it is not clear if the seals are killed by the lions, or scavenged along the coast. The distribution of the prey of lions in the Namib Desert is generally sparse. These lions do feed on another carnivore, the black-backed jackal, which is relatively abundant there. The lions also feed on white-breasted and Cape cormorants, both of which are abundant in rookeries along the coast. Namib Desert lions therefore clearly show the high degree of food selection opportunism present in lions generally (Bridgeford 1985). The same is true for lions in Somalia which depend heavily for food upon the warthog *Phaecochoerus aethiopicus* which is locally abundant there (Fagotto 1985). In comparison, 85% of the food eaten by lions in the Serengeti are buffalo *Syncerus caffer*, wildebeest *Connochaetes taurinus* and Burchell's zebra (Saba 1979).

The Asiatic lion is equally opportunistic in its food selection. A comparison of the food used from 1986 to 1990, to that eaten in earlier times, shows increasing use of wild ungulates, especially chital *Axis axis* and sambar *Cervus unicolor* (Chellam and Johnsingh 1993). The availability of these species also increased over these years after livestock were removed from the Gir National Park. Before 1971, the lions in that region mainly depended upon livestock for food (Johnsingh and Chellam 1991).

Surplus killing is known in lions, both in Africa and in India (Pienaar 1969; Kruuk 1972a), especially if a herd is weak from drought (Pienaar 1969). Both the African and the Asiatic lion also on occasion deliberately eat grass (Schaller 1972; Johnsingh and Chellam 1991). Male lions in particular are known to be cannibalistic (Pienaar 1969).

Leopards. The leopard is opportunistic in its food choice, killing prey of specific size classes rather than specific species (Bothma and Le Riche 1984, 1986; Kruger 1988). Leopards also have a variable diet between individuals (Kruuk and Turner 1967). Because leopards occur over a wide geographic range, their diet is more varied than that of any other large felid (Sunquist and Sunquist 1989). The prey of leopards in the arid bushveld of South Africa usually weighs from 20 to 70 kg, although the impala *Aecpyceros melampus* there forms 88% of the leopard's prey, mainly because of its local abundance (Kruger 1988). In food choice, leopards are adaptable. They sometimes become habit-killers by taking only or mainly a single food resource (Turner and Watson 1964). For example, in the Kafue National Park, a leopard which was re-located from a tempory island in Lake Kariba in Zambia, where it had become entirely adapted to a fish diet, continued to catch *Clarius* sp. after its relocation to the park (Mitchell et al. 1965). Leopards are also cannibalistic, but rarely so (Pienaar 1969).

Local prey abundance appears to be a common factor in the selection of food by leopards elsewhere too. In The Matobo National Park, for example, the hyraxes *Procavia capensis* and *Heterohyrax brucei* are abundant, and form from 55 to 61% of the leopard's diet (Smith 1978). In the Serengeti, a leopard hunts mainly from good cover, hence its diet seldom includes small animals from the open plains (Kruuk and Turner 1967). In the southern Kalahari, on the other hand, there is no dominant food resource in terms of numbers and distribution, and the leopards there use a variety of prey. However, prey items weighing 70 kg or less are taken most often (Bothma and Le Riche 1984). In the arid Augrabies National Park, no clear single prey species preference by leopards is also evident, although a relatively high incidence of porcupines, duiker *Sylvicapra grimmia* and gemsbok calves *Oryx gazella* occur in the leopard's diet (Bothma and Le Riche 1994a).

Other carnivores, but especially the black-backed jackal, aardwolf and bat-eared fox, are hunted and eaten regularly by leopards in the Kalahari Gemsbok and Augrabies National Parks (Bothma and Le Riche 1984, 1994a). In the Ngorongoro Crater, leopards occasionally even prey on African wild dogs (Wright 1960).

In areas such as Matobo National Park (Grobler and Wilson 1972; Smith 1978) and Tsavo National Park (Hamilton 1976), rodents are often used as food by leopards. Elsewhere rodents are rarely used (Le Roux and Skinner 1989). In more arid areas such as the southern Kalahari, rodents do not form part of a leopard's diet (Bothma and Le Riche 1994a). However, it is not impossible that rodents will be used in years of abundance. Even the snow leopard *Uncia uncia* of the arid rain shadow area of the Annapurna Mountain Range, eats rodents fairly regularly (Oli et al. 1993). Birds are seldom eaten by leopards in the southern Kalahari. Even when eaten, only the ostrich *Struthio camelus* is involved (Bothma and Le Riche 1989). In the Tsavo National Park, however, phasianid birds especially are eaten by leopards (Hamilton 1976). Reptiles (Hamilton 1976; Le Roux and Skinner 1989), invertebrates (Fey 1964; Ewer 1973; Hamilton 1976), fair quantities of grass (Smith 1978) and wild fruits (Bothma and Le Riche 1994a) are also eaten by leopards, but rarely so. Leopards also readily eat putrid meat (Smith 1978; Scott 1985). As a consequence, leopards can be trapped with such bait. Leopards will also scavenge food from other carnivores, and feed on animals that have died a natural death (Kruger 1988). In the Judean Desert the leopards are relatively small (males: 30 kg; females: 20 kg) in size, and their limited food resource leads to a diet of 63% ibex *Capra ibex* and 37% rock hyrax and porcupines *Hystrix africaeaustralis* (Skinner and Ilani 1979).

Cheetahs. This cat eats different prey species in different localities, but in the Serengeti young prey are taken in preference to adults (Caro 1994). There is also no clear evidence in the Serengeti that when adult prey are taken, old or sick animals are deliberately selected (FitzGibbon and Fanshawe 1989). In the Nairobi National Park, cheetahs use 25 types of prey, ranging from hares and warthog piglets to adult wildebeest and zebras. Specialization on particular prey by individual cheetahs occurs (Eaton 1970b). In contrast, cheetahs in the southern Kalahari mainly hunt springbok, but they also often hunt ostriches and duikers, and to a lesser extent five other prey animals. Attempts to hunt bat-eared foxes, black-backed jackals and the fleet-footed red hartebeest *Alcelaphus buselaphus* calves are invariably unsuccessful (Labuschagne 1979). In the Etosha National Park, cheetahs use the springbok as an important source of food (Phillips 1993). Although Bertram (1979) considers the cheetah to be the only large African carnivore that does not scavenge, this has been reported in isolated cases by Caro (1982) in the Serengeti, and by Pienaar (1969) in the Kruger National Park. Cannibalism by cheetahs also occurs (Pienaar 1969), but rarely so, and then probably only as a result of territorial disputes (Hunter and Skinner 1995).

Other Cats. The caracal is an opportunistic feeder, but it usually takes smaller mammals as food (Grobler et al. 1984). Caracal in the Postberg Private Nature Reserve make use of small rodents as their main source of food, concentrating heavily on the bush Karoo rat *Otomys unisulcatus* and the striped field mouse *Rhabdomys pumilio*. Also often eaten, however, is the steenbok, duiker, springbok, rock hyrax, hares *Lepus* sp., rodent moles (Bathyergidae) and smaller carnivores such as the otter *Atilax paludinosus*, Cape grey mongoose *Galerella pulverulenta*, polecat *Ictonyx striatus* and some birds (Avenant 1993). Some of these prey are taken abundantly only on a seasonal basis. In the Mountain Zebra National Park, the rock hyrax and mountain reedbuck *Redunca fulvorufula* are the main prey of caracals, followed by a variety of other mammals, birds, reptiles and invertebrates. However, on neighbouring farmland, rock hyraxes and rodents are the main food of caracal. This diet differs significantly from that of caracals in the adjacent national park (Moolman 1986).

In the mountainous regions of the Western Cape Province of South Africa, rock hyraxes, small antelope and vlei rats *Otomys irroratus* are also the main prey of caracals (Norton et al. 1986). In the Karoo National Park, the prey base available consists of fewer and larger mammals. Consequently, rodents form the main prey of the caracal there, with some use of the rock hyrax and the grey rhebuck *Pelea capreolus* (Palmer and Fairall 1988). In the Bedford farming area of the Karoo region, caracals switch to an almost exclusive diet of rock hyraxes following the elimination of other prey (Pringle and Pringle 1979). In Botswana, murids are the food most often eaten, but other animals such as birds, antelopes as large as adult springbok, hares, springhares and reptiles are also eaten (Smithers 1971). The caracal occasionally also scavenges food (Skinner 1979).

The nocturnal African wild cat in the Karoo National Park eats rodents mainly, with birds and invertebrates being taken occasionally (Palmer and Fairall 1988). Springhares and hares are the largest prey of this cat, while murids form 50% of all wild cat food in Botswana. Insects, reptiles and birds are also eaten. The black-footed cat eats insects, spiders, frogs, lizards, small birds and murids (Smithers 1971; Grobler et al. 1984). The sand cat *Felis margarita* digs for fossorial rodents, one of its main food items. The diet also includes birds, reptiles and arthropods (Nowell and Jackson 1996). The sand cat has a reputation as an adept snake hunter, particularly of horned and sand vipers (Dragesco-Joffé 1993; Nowell and Jackson 1996).

3.5.2 Hunting

Lions. The cats differ widely in their hunting methods. This in part is the result of habitat diversity, prey availability and the size and social structure of the species involved. The African lion pride often hunts as a group, with individuals spreading out to take different routes to the intended prey (Stander 1992a; Fig. 3.3). Lions hunting as a pride in the Savuti Marsh will deliberately hunt any small animal cooperatively. In doing so, however, the individual making the kill generally also eats it without sharing it (McBride 1990). Group hunting occurs mostly at night and is believed to require less cover than solitary hunting (Bertram 1979). However, the success or failure of a given hunt depends as much upon the response of the prey to the hunting lion or lions as it does upon their hunting prowess. As long as hunting lions are visible to them, prey are quite casual in their response. Therefore, a lion must seek every possible advantage to kill its intended prey. Lions are believed by some authors seldom to hunt by using the prevailing wind direction, but the height and density of vegetation influence the ease of stalking. Sight and smell are used extensively in stalking (Schaller 1972). Nonetheless in the Etosha National Park, lions do hunt significantly more often and more successfully downwind (Stander 1992b).

Female lions hunt proportionately more often than males. Opportunistic hunts have the highest success rate (61%) in the Serengeti ecosystem. Group hunting by lions increases the chances of success, especially against larger prey (Schaller 1972; Caraco and Wolf 1975), although males are less likely to hunt in groups than females. Both male and female lions in the Serengeti often refrain from participating in a hunt depending upon the type of prey, and ease of killing it (Scheel and Packer 1991). For small prey, the optimum hunting group size and the maximum allowable group size, based upon energetic balances, are the same (Caraco and Wolf 1975). In exceptionally dry years in the southern Kalahari, a larger pride kills much more frequently than a smaller one (Eloff 1973a).

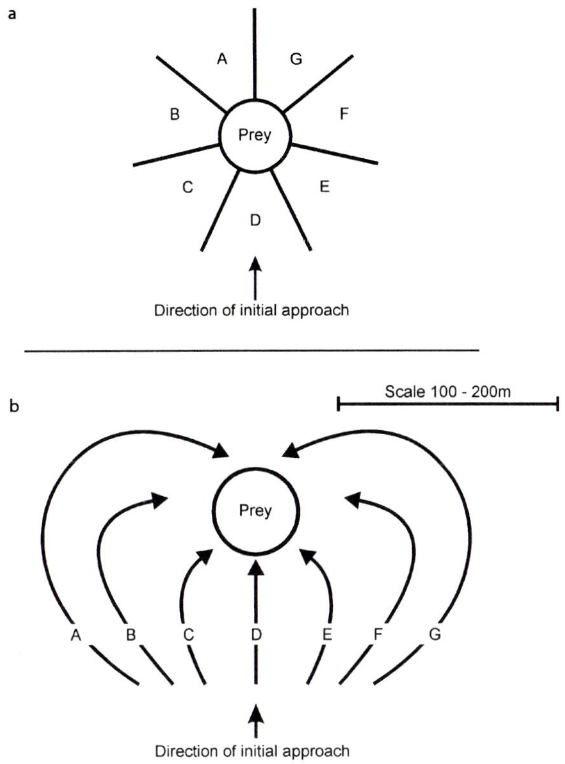

Fig. 3.3 a, b. Typical stalking routes **b** of individual lions (A to G) in the Etosha National Park in Namibia, intended to surround **a** the prey for the final kill (Stander 1992a)

McBride (1990) believes that the lions of the Savuti, where 90% of all hunts occur at night, do so, as they do elsewhere, because of high heat stress of the African day. Actual hunting by a lion begins with a systematic search, not directed towards a particular animal but intended to scan the environment generally. This is followed by a stalk of varying distance, using as much cover as possible. Depending upon the terrain and the behaviour of the prey, the lion gets close to its intended prey, making a high-speed, final charge towards the victim. Chases are generally believed to seldom last for more than 100 m or so (Kruuk and Turner 1967). However, McBride (1990) believes that the African lion is capable of sustained bursts of speed over distances of as much as 1 km. Stalking can occur in bouts, with the lion stopping at times, or it can be a slow, sinuous and continuous action (Kruuk and Turner 1967). In the Kafue National Park, hunting conditions for lions are easy and physical strength more than any other factor will determine the success of a hunt. For example, the lechwe *Kobus leche* is hunted with great difficulty there, whereas the buffalo is a major buffer species in the dry season, easing the predation pressure on other prey species (Mitchell et al. 1965). Prey availability influ-

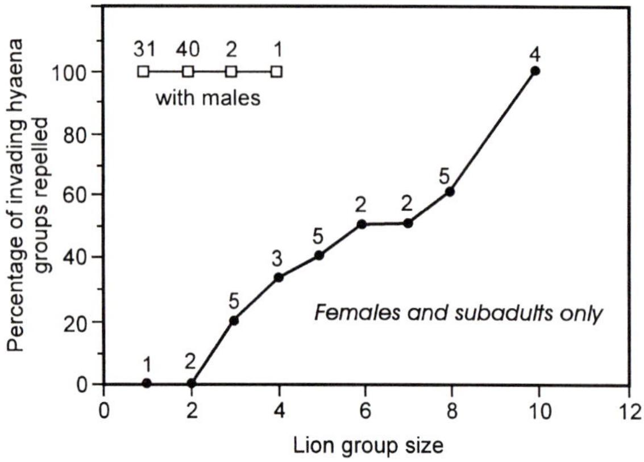

Fig. 3.4. The proportion of kills at which lions of different group sizes were able to repel spotted hyaenas of different group sizes from their kills in the Savuti Marsh of the Chobe National Park in Botswana. Sample sizes appear above the data points (Cooper 1991)

ences the focus of a given hunt. The relatively prey-poor southern Kalahari, for example, is the only area where lions consistently hunt small mammals (Smuts 1979). In the southern Kalahari, lions mostly (69%) kill small prey or the young or subadult segment of the available prey population (Eloff 1973a). Lions of the Savuti also mostly (62%) kill young or subadult prey (McBride 1990).

In the Savuti, there is also preliminary evidence that kleptoparasitism by spotted hyaenas may influence the optimal hunting group size in lions (Fig. 3.4). The Savuti lion population suffers from a shortage of adult males. In their absence (44% of feeding groups have no adult male close by), spotted hyaenas can drive female and subadult males from their kills, provided that they outnumber such lions by a ratio of 4:1. The larger the lion hunting group, the better it defends the kill. The lions there lose almost 20% of their food to spotted hyaenas, but these lions are often already satiated by the time that the hyaenas acquire the kill. Also, these lions often hunt in groups large enough to defend their kills, even in the absence of a large adult male. The hyaenas therefore do not cause an immediate need, but they do constitute a constant energy drain on the Savuti lions by forcing them to hunt more frequently than they would have done in the absence of this interaction (Cooper 1991).

In the arid and relatively open southern Kalahari, hunting lions take advantage of every possible cover, even small shrubs and tufts of grass. To make use of the relatively abundant, but potentially lethal, gemsbok (Fig. 3.5) as a prey resource (Eloff 1973a), the Kalahari lion has altered the focus of its attack by approaching gemsbok from behind and attacking the lumbar region, instead of using the normal throat and neck attack. The lion then jumps onto the haunches of the gemsbok out of reach of the lethal horns, and breaks its prey's back by snapping the spinal chord, before seizing its throat and crushing the neck. The back is not broken by the impact of the attack, but by a sharp upward jerk with the jaws, focusing on what seems to be a weak link

Fig. 3.5. In the southern Kalahari, the gemsbok is a potentially lethal prey of large carnivores

in the anatomy of this and many other ungulates in which vertebral anatomy has sacrificed stability for mobility. A gemsbok's horns are lethal weapons and cases are known in which an attacking lion has been gored and killed by a gemsbok (Eloff 1964, 1973a).

Lions of the southern Kalahari also show clear, cooperative hunting as is found elsewhere, with the females being especially important. However, the method used to kill a gemsbok indicates a prey-specific, learned hunting technique resulting from the smaller abundance and variety of prey in the southern Kalahari. In contrast, the lions of the Etosha National Park also kill gemsbok, but do so with the customary throat and neck bite. This difference probably reflects the greater abundance and variety of prey in Etosha compared with the southern Kalahari (Eloff 1973a).

In the Etosha National Park, group hunting by lions involves the circling of prey by some females of the pride (Fig. 3.3), while others wait for the prey to move to them. The circling lions frequently initiate the attack, while the waiting animals move relatively short distances and usually capture the fleeing prey. In Etosha, a lioness in a pride usually occupies the same position in the hunting formation repeatedly. Hunts where most of the lionesses present occupy their preferred position in the hunting formation have a higher probability of success than others. However, hunting behaviour is flexible, varying from group to group (Stander 1992a,b).

Cooperative hunts by lion groups are more common where prey are difficult to catch. On the open plains of the Etosha National Park, lions have low overall capture success rates (15% for all prey species in both the wet and dry seasons). Single lionesses have the lowest success rate (2.3%), while cooperative hunts have a high success rate (27%). During successful, cooperative hunts most prey are caught from an ambush. The individual that captures the prey, always also kills it (Stander 1992a,b).

Cooperative or group hunting by the lions of Etosha clearly benefits the individual member's efficiency and per capita food intake, especially during the dry season when prey is scarce (Stander 1992a). During this season, in fact, group hunting is essential for lions to gain sufficient food. A pair of lions is the optimum group size for hunting. In the wet season, hunting groups of all sizes obtain more than their daily food requirements. Etosha lions rarely scavenge as they seem to obtain sufficient food by hunting at most times (Stander 1992b).

Asiatic lions live in conspicious groups in open habitat. Females prefer to hunt near water where natural prey congregate, avoiding livestock which are usually associated with human presence. Male lions frequent hilltops in the wet season, probably because the cool breezes there reduce the impact of insects. Unlike African lions, male and female Asiatic lions seldom congregate or even hunt together. The Asiatic male lion will only join a female to mate or to feed at a large kill (Johnsingh and Chellam 1991).

Leopards. The leopard usually hunts and kills in one continuous action. It hunts from cover more so than any other predator (Kruuk and Turner 1967). A stalking leopard uses every bit of available cover (Eltringham 1979). In the Serengeti ecosystem, leopards may hunt from an ambush or by stalking, rushing at the prey in a high-speed dash for the kill (Kruuk and Turner 1967; Bailey 1993). Attacks are apparently never launched from trees (Le Roux and Skinner 1989; Bailey 1993). Leopards hunt opportunistically, mainly at night, dawn or dusk, but they will hunt in broad daylight if the opportunity arises. When doing so, they often watch their intended prey for some time before starting to hunt (Scott 1985; Bailey 1993).

In arid areas with limited resources, adaptive hunting behaviour has an important survival value for a large predator such as the leopard (Bothma and Le Riche 1989). In the southern Kalahari, leopards hunt opportunistically and are therefore number maximizers as opposed to the leopards of more mesic areas which are energy maximizers because of greater prey abundance (Bothma et al. 1997; Karanth and Sunquist 1995). As in other places where prey is scarce (Turnbull-Kemp 1967), southern Kalahari leopards only hunt in any specific part of their range once every four to six days, thereby giving the prey in a given area time to settle down again before their return (Bothma and Le Riche 1984).

When hunting, the leopards of the southern Kalahari use a flexible approach based upon the vigilance, vulnerability and aggression of their prey. Vigilant prey such as the ostrich, and potentially dangerous or lethal prey such as a gemsbok calf attended by its mother (Eloff 1964, 1973a), are stalked with great care over distances of several kilometres to position the leopard optimally for the shortest possible final rush to the kill. Should the potential danger persist after the kill, such as when a gemsbok calf is closely attended by its mother who will often defend the carcass of the calf, the leopard will retreat to some suitable cover. From there it will watch its prey until the calf has been abandoned by the cow. A gemsbok calf attended by its mother is probably the most dangerous of a Kalahari leopard's adversaries. The leopards there have clearly developed a hunting strategy to cope with this danger when utilising this relatively abundant prey resource (Bothma and Le Riche 1989).

However, it is not only for lethal and vigilant prey that leopards in the southern Kalahari adapt their hunting tactics. Every type of prey animal has its own habits and individual characteristics such as wariness, aggression and fleet-footedness to which leopards adopt different and prey-specific hunting methods. While some prey are hunted in the same way, others may be hunted differently. Male leopards, being considerably larger than females, at times even hunt the same type of prey in a different way than that adopted by the females. However, other prey are hunted in the same way by both males and females (Bothma et al. 1997).

Southern Kalahari leopards only stalk their prey in 35% of cases, and the full stalk,chase and kill sequence is used in only 29% of all the kills. Where stalking does occur, optimal positioning, long stalking distances and excessive care are involved in 24% of cases. One gemsbok calf attended by its mother was stalked for 3.4 km, an intervening dune ridge being used as cover. In this particular instance the leopard looked over the dune ridge twice during stalking to determine the position of the prey before rushing from not more than 10 m and killing it (Bothma and Le Riche 1989). Although Kalahari leopards often chase black-backed jackals and kill them, these jackals are seldom stalked deliberately. Of 228 stalks recorded for male leopards, none targeted black-backed jackals. Female leopards only once did so in 146 stalks. This probably happens because black-backed jackals often follow leopards when they hunt (Sect. 3.7). In such instances, the leopard does not need to stalk the jackal because both are already aware of each other. Instead, the leopard simply chases and occasionally kills the jackal (Bothma, unpubl.).

The mean stalking distance for all prey by leopards in the southern Kalahari is 142.7 m (n = 228) for males and 129.4 m (n = 80) for females. In comparison, the mean stalking distance for gemsbok is 365.6 m (n = 57) in males and 371.9 m (n = 9) in females. When adult male leopards stalk prey, only 10.9% of all the stalks end in a kill. When adult female leopards do so, only 9.5% of all stalks end in a kill (Bothma et al. 1997). This shows the high level of energy expenditure required for southern Kalahari leopards to obtain their food.

Of all stalks by male leopards in the southern Kalahari, 24% lead to a chase and 30% do so in the case of females. Small prey are seldom stalked, particularly by male leopards, and are usually merely pounced upon, often while lying down, presumably resting (Bothma and Le Riche 1984, 1986). When prey is chased, the distances involved vary according to prey type. Porcupines go into a stationary defence mode with raised quills and are only chased for a mean distance of 14.2 m (n = 13) by male leopards. They are hunted infrequently by females. In comparison, the mean chase distance of all prey is 48.5 m (n = 146) for male leopards and 36.8 m (n = 61) for females. For all prey, 14.6% (n = 137) of chases by males end in kills, 27.9% (n = 61) do so in females (Bothma et al. 1997).

Although the southern Kalahari is an arid, open landscape of shrub and woodland savanna, hunting cover is not a limiting factor for leopards in this region. What cover there is, is sufficient to allow stalking leopards to get within striking distance of their prey. A stalking leopard can conceal itself behind a shrub or tuft of grass as low as 200 mm. Moreover, although it is an open environment, the density of 58.5 woody plants per ha (shrubs and trees) in the interior tree savanna habitat is sufficient to allow successful stalking (Bothma et al. 1994).

Hunger, which is one of the basic behavioural motivations for hunting by various types of predator (Schaller 1972; Leyhausen 1979), does not seem to influence the stalking or chase distances of southern Kalahari leopards (Bothma and Le Riche 1989,1990). However, as hunger levels increase, southern Kalahari leopards do increase their hunting intensity. This, however, manifests itself by increased distances moved per 24-h period rather than by increased frequency of hunts. The evidence suggests that southern Kalahari leopards, whose prey is scarce, are already utilizing their available prey resource maximally by hunting every available item in any given hunt. With the number of hunts per km remaining constant, a leopard can only increase the overall number of hunts per 24-h period by moving increasingly longer distances as its hunger grows. A recently satiated leopard moves a mean of 10.2 km per 24-h period. In comparison, a leopard that has gone without food for one day, moves a mean of 14.2 km over the same period, one that has been without food for two consecutive days, moves a mean of 20 km, and one that has had no food for three or more consecutive days, moves a mean distance of 21.8 km in 24 h. This difference is significant. It increases the total number of prey contacts as hunger increases without changing the rate of prey contacts per kilometre moved (Bothma and Le Riche 1990).

Cheetahs. The hunting success of the cheetah varies with habitat, prey type, sex and age, herd size, cheetah group size and the individual cheetah's or group's hunting experience. The cheetah depends mostly upon vision and speed and not smell to hunt (Eaton 1970a). When chasing prey, a cheetah is said to be able to reach a speed of up to 110 km/h (Grobler et al. 1984). In reality, such speeds are rare in the field with all its obstructions, and prey jinking and swerving to avoid capture. In the wild, 78 chases measured by Frame and

Frame (1981) yielded a top chase speed of 87 km/h in cheetahs. Antelopes can reach top speeds equal to that (Garland 1983; Nowell and Jackson 1996). A hunting cheetah watches its prey intently, stalking with a slow walk and not a crouch as other large cats do. No division of labour occurs in a group of cheetahs, and hunting is the same when the cheetahs are in a group as when they are single. The chase, once begun, ends when the prey escapes or is knocked over and killed (Eaton 1970c). The cheetah is a sprinter rather than a courser (Nowell and Jackson 1996). Male cheetahs that live in groups also hunt together. The weight of prey taken increases with the size of the hunting group (Caro 1994), as in spotted hyaenas. Cheetahs in the Nairobi National Park select juvenile or female prey (Eaton 1970b).

In the southern Kalahari, single cheetahs successfully kill in 32.4% of all hunting attempts when single, 25% when hunting in pairs of adults and 71.4% when a female with cubs younger than 12 months old is hunting. In one instance, a female with young cubs is known to have hunted nine times in 22 days, being successful in 77.7% of these attempts. As the southern Kalahari is poor in prey, there is no selection for prey type or attributes, and all potential prey are hunted opportunistically with no obvious preferences being shown (Labuschagne 1979).

Cheetahs in the southern Kalahari move a mean distance of 10.2 km between unsuccessful hunts and 30.7 km between successful ones. This translates into a successful hunt every 2.5 days or a mean hunting rate of once in every 7.6 km travelled. Single cheetahs there successfully hunt once every 24.4 km and adult pairs once every 39.9 km. In terms of energy gained, however, pairs are more successful because they hunt and kill more larger prey than single cheetahs (Labuschagne 1979).

The difference in the size of prey selected by single, adult cheetahs as opposed to pairs in the Kalahari is evident from the following: in increasing order of size, the prey animals killed by single cheetahs were, Cape hares *Lepus capensis* in 25% of all kills, springhares 33.3%, steenbok 12.5%, duikers 8.3%, springbok 8.3%, and ostriches 12.5%, while gemsbok adults and calves were never killed. In contrast, adult cheetah pairs killed Cape hares in only 5% of all cases, they never killed springhares and duikers, killed steenbok in 10% of all cases, frequently killed springbok (45%) and ostriches (30%), and they even killed adult gemsbok and calves (10%)(Labuschagne 1979).

In the southern Kalahari, cheetahs find prey in three ways: by chance, by a purposeful search using vantage points such as trees or termite mounds, or by ambushing prey from concealment. The prey is located mainly by sight. Successful kills involve a longer chase distance (mean: 218 m) than unsuccessful ones (mean: 122 m). The maximum distance recorded is 650 m for an unsuccessful chase of an ostrich. The maximum successful chase distance recorded is 550 m for an adult springbok. A close approach before launching the attack apparently does not ensure greater chances of hunting success (Labuschagne 1979).

As for southern Kalahari leopards (Bothma et al. 1997), steenbok (Fig. 3.6.) are hunted frequently by cheetahs, but are equally infrequently killed thanks to their vigilance and fleet-footedness (Labuschagne 1979). Similarly, gemsbok are aggressive especially when their calves are threatened, although they are frequently killed by leopards (Bothma et al. 1997) and by cheetahs (Labuschagne 1979). Cheetahs do chase smaller carnivores such as black-backed jackals and bat-eared foxes in the southern Kalahari, but this seems to be without any real hunting intent (Labuschagne 1979).

Other Cats. Few data exist on hunting by other cats in arid areas. However, caracal in the Mountain Zebra National Park are known to hunt by stalking or from an ambush (Moolman 1986).

3.5.3 Killing

The canines of felids display relatively fewer features (pits) of microwear than canids or hyaenids, suggesting that felids deliver deep, strong bites when killing their prey. These bites result in less abrasion of the canines than in other large carnivores (Anyonge 1996).

The African lion is well equipped for grasping and killing its prey with its claws and teeth, despite lacking great speed (Schaller 1972). The prey is usually killed with a throat or muzzle bite leading to suffocation (Kruuk

Fig. 3.6. The steenbok is hunted often, but infrequently killed, by cheetahs and leopards in many parts of southern Africa

and Turner 1967) or a broken neck (Eloff 1973 a, c). However, the killing method is adaptable and, as already said, the lion in the southern Kalahari kills gemsbok by jumping onto the victim's haunches and breaking its back at the lumbo-sacral joint. It seldom breaks the neck of other animals, but the prey taken in the southern Kalahari is generally small and easily killed with a few bites (Eloff 1964, 1973a, 1984). Compared with the leopard there, the lion in the southern Kalahari has a high success rate as it kills 38.5% of all prey hunted. However, kill rates vary with the prey selected. For example, lions in the southern Kalahari kill only 13.6% of all aardvark hunted, 15% of all adult gemsbok hunted, but 85% of all porcupines hunted (Eloff 1984).

The leopard commonly kills its prey with a throat or neck bite, or one to the base of the skull (Kruuk and Turner 1967; Smith 1978; Bothma and Le Riche 1984, 1986; Scott 1985). Leopards are proficient hunters and can kill prey three to four times their own weight (Bailey 1993). Southern Kalahari leopards most often kill early in their daily movements. This coincides with the early evening (Bothma and Le Riche 1986). In the Serengeti ecosystem, leopards kill 11% of all prey hunted (Schaller 1972). In contrast, male Kalahari leopards kill 18.5% of all prey hunted and females 21.8%. However, prey in the southern Kalahari are not killed in proportion to the frequency with which they are hunted. For example, the steenbok, an abundant, fleet-footed and vigilant small Kalahari antelope is the focus of 22.2% (n = 410) of all hunts by male and female Kalahari leopards, but it only represents 7.5% (n = 80) of all prey killed by them (Bothma et al. 1997). In the Matobo National Park, the steenbok is also common, but it is infrequently killed by leopards, being found in only 2% of all scats (Grobler and Wilson 1972). In the arid bushveld of South Africa, steenbok are also abundant but are never killed by leopards there (Kruger 1988). In the southern Kalahari, the steenbok is by far the prey hunted most unsuccessfully, requiring 27.5 hunting attemps for every successful kill by male leopards, and 9 attempts by females. In contrast, it requires a mean of 5.4 and 4.6 hunts respectively per successful kill for all prey (Bothma et al. 1997).

The cheetah commonly kills larger prey by suffocation through a throat or muzzle bite (Kruuk and Turner 1967) after a high-speed chase. This chase usually ends when the cheetah slaps the hindlegs of its prey to unbalance and overpower it (Eaton 1970c; Phillips 1993). Smaller prey, however, are simply pounced upon, pinned to the ground and killed with a single bite (Eaton 1970c; Labuschagne 1979). In the southern Kalahari, suffocation of larger prey by a cheetah may take up to 10 min. Moreover, the prey sometimes recovers from the initial suffocation attempt, usually only to be suffocated again (Labuschagne 1979).

A caracal catches prey with a sideways swipe of the front paws (Smithers 1971) and usually bites its prey in the throat or at times in the nape of the neck. The lower incisors of a caracal are reduced. Coupled with its large diastema, this allows the canine teeth to penetrate the prey almost to their full length (Pringle and Pringle 1979; Moolman 1986).

Surplus killing as defined elsewhere, occurs occasionally in felids but it seems to be rare. Nevertheless, in the Gir Forest of India, a single leopard once killed 17 of a group of 60 domestic goats that went astray in a monsoon (Kruuk 1972a). In the Mountain Zebra National Park a caracal once killed small livestock en masse, but no natural prey are known to have been killed in a similar way (Moolman 1986).

3.5.4 Feeding

Lions. The carcasses of killed prey are eviscerated before feeding starts. In feeding, lions eat some bones, but they eat muscle mainly, using their front paws in 35% of all cases to aid in feeding (Van Valkenburgh 1996). When feeding at a kill, lions are quite socially organized. The larger males are dominant over the females, but there is considerable squabbling (Bertram 1979). Lions consume large prey as completely as possible. In the Serengeti ecosystem, lions sometimes cover the disembowelled viscera of their prey with vegetation or soil (Schaller 1972). However, in the southern Kalahari, they regularly cover these remains by scraping sand over them. The viscera are often removed from the carcass and is then dragged some distance away before being covered. The remainder of the food may then be moved a short distance to suitable cover before feeding starts, or it may be eaten where it is (Eloff 1973a). Lions in the Serengeti pluck out the long hair of furry prey with their incisors before beginning to feed (Schaller 1972). When a porcupine is killed in the southern Kalahari, it is carefully dequilled to a fair extent before being eaten. Nevertheless, southern Kalahari lions may at times sustain fatal injuries from quills when attacking or eating porcupines (Eloff 1973a). Usually, however, through its feeding habits and daily pattern of life, the lion of the southern Kalahari is able to meet its habitat's greatest challenge: that of lack of water and excessive heat (Eloff 1973a).

Near the coast in the Namib Desert, lions drag carcasses considerable distances from the kill site (mean: 1100 m; range 100 to 2600 m) before starting to feed. Dragging carcasses from a kill site before feeding is common felid behaviour, but in the Namib Desert the carcasses are usually dragged by lions away from exposed, windy beaches to the shelter of vegetated hummocks, or into the reedbeds of the dry river courses. Away from the coast, the carcasses are usually dragged to the shelter of some isolated small tree or bush. This action also hides the carcass from possible scavenging spotted hyaenas (Bridgeford 1985).

The mean daily food intake of an African lion indicates an optimum lion group size for the ecological conditions of a given area. Prey biomass and lion feeding group size define the limits of acceptable energetic expenditure or gain, (Caraco and Wolf 1975). In the Kruger National Park, lions feed only once every four days (range 1 to 13 days), while 47.1% of all lion stomachs are empty at any given time (Bryden 1978).

Leopards. Before feeding, the leopard usually also drags the carcass to some form of acceptable cover (Smith 1978; Bothma and Le Riche 1984, 1986). In areas of high terrestrial predator density, leopards often store their food in a tree to avoid it being scavenged by other predators, especially spotted hyaenas (Scott 1985). For example, 31% af all kills by leopards in the arid bushveld of South Africa are stored in trees (Kruger 1988). However, actual interference by scavengers is not the only cause of leopards storing prey in trees as they often do this, possibly as a preventive move or for some other reason (Kruger 1988). In the southern Kalahari, where large predators are not abundant, leopards only store 17% of their kills in trees before feeding (Bothma and Le Riche 1984). Especially, but not necessarily only when eating furry prey such as young antelope, hares or other carnivores, leopards also use their incisor teeth to pluck out small bundles of hair so as to gain access to the meat, dropping the hair in a neat pile (Bothma and Le Riche 1984; Scott 1985; Kruger 1988). Porcupines are dequilled as lions do before being eaten (Bothma and Le Riche 1986). Most prey are eviscerated en route to the feeding site or at the kill site (Smith 1978). Only female leopards in the southern Kalahari, however, are known to cover the eviscerated remains of their prey with vegetation, and they only do so when the kill is made close to small cubs (Bothma, unpubl.). However, in areas of higher scavenger density such as the Matobo National Park, such remains are covered more frequently, but not regularly (Smith 1978). In the arid bushveld of South Africa, leopards never cover the intestines of their prey with sand or vegetation (Kruger 1988) despite the abundance of scavengers.

In the Matobo National Park, kills made in the wet season are usually dragged to higher ground for a mean distance of 260 m. In the dry season they are only dragged a mean distance of 120 m (Smith 1978). In the southern Kalahari, male leopards drag their prey longer distances away from the kill site (mean: 410 m) than females (mean: 57 m). The reason for this difference is unknown (Bothma and Le Riche 1984). Leopards in the southern Kalahari occasionally drag or carry their prey considerable distances before starting to feed. In the Matobo National Park, a leopard once carried a rock hyrax 1.6 km before starting to feed (Smith 1978). In the southern Kalahari, a male leopard once carried a bat-eared fox 4.9 km before doing the same (Bothma and Le Riche 1984). There is no apparent preference for the type of vegetation used by leopards for feeding (Kruger 1988).

Usually a leopard starts to feed on the buttocks around the anus of its prey, but occasionally it starts on the chest or shoulders (Smith 1978; Scott 1985; Bothma and Le Riche 1986; Kruger 1988). The tongue of a leopard is covered with hair-like papillae which are used to rasp meat particles from bone (Scott 1985). The muzzle, part of the cranium, feet and viscera of prey are seldom eaten (Smith 1978; Bothma and Le Riche 1986).

Leopards will sometimes feed on a kill for several days. In the southern Kalahari, only a few kills are big enough, however, to satisfy a leopard for longer than a day. The maximum feeding duration recorded there is four days for a female that had killed a large aardvark. Male leopards in the southern Kalahari consume approximately 3.5 kg of meat per day; females with cubs 4.9 kg, but the males only kill once every 3 days as opposed to once every 1.5 days by females with cubs (Bothma and Le Riche 1984). In the Matobo National Park, a leopard once fed on a yearling blue wildebeest for six days (Smith 1978). Leopards which are disturbed at a kill in the southern Kalahari usually return to it later (Bothma and Le Riche 1993).

Cheetahs. Cheetahs only feed on food killed by themselves (Bertram 1979). When doing so, they feed as rapidly as possible to avoid competition from scavengers, although they may stay with a kill for up to 11 hours. When feeding, cheetahs in the Etosha National Park eat all the bones, except the skull, of prey weighing less than 10 kg. Even young cheetahs six to 16 months old have no difficulty in crushing and consuming most bones of a kill (Phillips 1993). Cheetahs do not use their front paws to aid in feeding as lions do (Van Valkenburgh 1996). Cheetahs feeding in groups are less likely to be harassed by scavenging spotted hyaenas in the Serengeti ecosystem than individuals feeding alone. A male cheetah will also occasionally scavenge food from a female there (Caro 1994). In the Etosha National Park, feeding cheetahs seldom pay attention to scavenging jackals, but may lose much or all of their prey to other scavengers. The cheetahs there feed in bouts lasting a mean of 82 min, with usually less than 15 min between bouts. If undisturbed, large prey such as an adult springbok are fed on for up to six feeding bouts. Each feeding bout lasts several hours. Adult cheetahs in the Etosha National Park require 3 to 4 kg of food per day to maintain good health (Phillips 1993).

In the southern Kalahari, cheetahs usually carry or drag their prey to shade before beginning to feed. The rump or ribs are eaten first, but little skin is eaten (Labuschagne 1979). When several cheetahs feed at one kill, they position themselves in the shape of a star around the kill, sometimes threatening each other as the food supply becomes reduced (Labuschagne 1979).

Other Cats. Before feeding, the caracal plays with its food (Smithers 1971). In the Mountain Zebra National Park, the prey is usually jerked about with the paws for five to 20 min before feeding begins: the reason for this is unknown. The caracal usually starts to eat the ribs and the breastbone, but it never touches the stomach or intestines (Smithers 1971; Pringle and Pringle 1979; Moolman 1986). The hair of certain prey is not removed before feeding as it is in the case of the leopard (Moolman 1986). Although the caracal does not eat carrion (Pringle and Pringle 1979), it will leave and then return to its own fresh kill for a second feeding bout later (Moolman 1986). Caracals do not cache or bury unconsumed parts of their prey, but occasionally a kill may be dragged into the fork of a tree for feeding (Pringle and Pringle 1979).

3.5.5
Food Passage and Defecation

Data on food passage and defecation rates in felids in the wild are scarce, but it is known that lions in the Kruger National Park have a slower rate of digestion of food, and that they also eat less frequently, than do spotted hyaenas (Smuts 1979). In the southern Kalahari, most remains of the food of leopards appear in the scat on the first day following ingestion, but some appear as much as five days later. There is no differential passage rate for different types of prey (Bothma and Le Riche 1994a). Leopards that feed on large prey, deposit most of their faeces near the feeding site (Eisenberg and Lockhart 1972).

The mean interval between successive defecations by a leopard in the southern Kalahari is 0.6 days for males and 1.2 days for females. The difference is significant, but the reason is unknown. It contradicts the general belief (McNab 1989) that carnivores with smaller body sizes (here the females) have higher basal metabolic rates (and hence defecate more frequently) than those with larger body sizes (here the males). The distance travelled between successive defecations by male (mean: 12.4 km) and female (mean: 21 km) leopards is random, but there is a significant difference between the sexes. Seasonal climatic differences do not appear to influence defecation frequency as a measure of metabolic rate in leopards in the Kalahari (Bothma and Le Riche 1994a).

3.5.6
Water Use

Like most carnivores, felids are independent of free water, but will utilize such water when available (Grobler et al. 1984). In the Umfolozi Game Reserve, lions do not derive any significant fraction of their water intake in cool weather from drinking. However, in hot weather about 50% of their water intake is obtained from drinking water (Green et al. 1984). Lions in the northern Sudan have no access to surface water and must obtain moisture from their prey (Schaller 1972). In the southern Kalahari, lions are found up to 250 km away from the nearest open water and are independent of drinking water (Eloff 1973a). While southern Kalahari lions drink regularly when water is available, they can be completely independent of water under extreme desert conditions, obtaining sufficient moisture from the body fluids of their prey, and from wild fruits such as tsama melons when available. Loss of water in southern Kalahari lions is further reduced by long rests in the shade during the hot hours of the day. When it rains, these lions lap up water forming in pools or lick the moisture from each other's pelage. Lionesses with cubs, however, drink water more frequently than other lions (Eloff 1973b).

Where available, leopards make use of caves as an important survival strategy to escape high daytime temperatures in hot climates, and to reduce water loss. In the Namib Desert, some caves may be used for many years by successive generations (Bailey 1993). Where caves are absent, as in the southern Kalahari, leopards rest underground by day in porcupine or aardvark burrows, or do so under the dense shade of vegetation such as the shepherd's tree (Bothma and Le Riche 1984, 1986). In desert areas, leopards are independent of free water but will drink it when it is available (Bailey 1993). Moisture is obtained through the body fluids of the prey, and by eating wild fruits when available (Bothma and Le Riche 1984, 1986). In the Masai Mara National Reserve, leopard cubs lick moisture from each other's fur when it rains (Scott 1985).

When water is available in the southern Kalahari, male leopards drink at a mean interval of 3.9 days, or once every 58 km moved. The male leopards move a maximum recorded distance of 113.4 km between drinks, while females with cubs seldom drink . However, they kill prey on average once every 1.5 days. On one occasion, an adult male leopard did not drink water for ten consecutive days in summer (Bothma and Le Riche 1986). Leopards lose little moisture through sweating, but pant rapidly as a cooling device (Scott 1985).

In the southern Kalahari, two adult male cheetahs are known to have moved 127 km in 12 days without drinking water. During this time they failed 15 times, but succeeded twice in killing prey. As with lions and leopards, wild fruits such as the tsama melon are used when available as a water source by cheetahs (Labuschagne 1979). No data on water use by other felids in arid areas are available.

3.6 Population Ecology

Little solid data regarding population ecology exist for felids in arid areas. What is known is fragmented and scattered in the literature. This area of study needs considerable scientific attention and is as yet a wide open field of study.

3.6.1 Lions

Mortality seldom targets adult African lions, but cubs are at risk. Most cubs die when still young. In the Serengeti ecosystem, most lions die from disease, starvation, abandonment, violent contacts with rival lions or other animals, or a from combination of these factors. Only 10% of all lions reach old age there (Schaller 1972). In a recent outbreak of canine distemper, 33% of an estimated 3000 lions in the Serengeti died (Packer and Kock 1995). Lions are immune to anthrax probably because the pathogens of this disease cannot flourish in the highly acidic digestive tract of lions, even when they feed on the carcasses of animals that have died of anthrax (Berry 1983). Healthy, adult lions seldom starve to death, but cubs often do. Abandonment of cubs is a major cause of death. In various areas of the Serengeti, more than 50% of all cubs die before reaching adulthood (Schaller 1972).

Of the lions in the southern Kalahari, 68% are adults and 32% cubs (Mills et al. 1978). However, cub mortality may be as high as 90% especially in years of extreme drought when prey is scarce (Eloff, pers. com.[2]). Females outnumber males by 1.8:1 in subadults and by 2.1:1 to 2.8:1 in adults (Eloff 1973a; Mills et al. 1978). Female lions in the Savuti Marsh outnumber males by a wide margin. There the sex ratio is 5.7 females per male, a ratio which has serious foraging implications for these lions as has been explained earlier (Cooper 1991). In more mesic regions of Africa, the sex ratio of adult lions is closer to unity. This is the case in the Selous Game Reserve where cubs form a smaller proportion (21%) of the population (Rodgers 1974). In the Nairobi National Park, adult female lions are also more abundant than adult males (Rudnai 1973), as they are in the Etosha National Park (Stander 1991). These variable sex ratios indicate that survival in African lions is dependent upon age and sex, but variable between populations.

Prey abundance is a major factor in lion population dynamics, especially in respect of mortality (Schaller 1972; Bertram 1979; Hanby et al. 1995). In the arid bushveld of South Africa, lions disperse in response to diminishing prey in times of drought. They use the giraffe as staple food because this herbivore is little affected by anything except prolonged drought (Kruger 1988). In the Serengeti, prey biomass doubled over a recent decade, leading to a 33% increase in the number of lion prides in the area. At the same time, the mean pride size increased from 15 to 19, while cub survival also increased (Hanby and Bygott 1979).

When culled, lion numbers are soon restored by infiltration from surrounding areas in the Kruger National Park, and by increased reproduction. Short-term culling of many lions over a large area creates many vacant areas that attract strange lions. These lions interact aggressively, probably suppressing the birth rate and cub survival in their own prides. A more gradual culling of a few lions at a time from neighbouring prides, or even the gradual removal of an entire pride over time, produces fewer vacant areas, a lower influx of new lions and more sustained reproduction. However, both methods are followed by a rapid recovery of the lion population in a given area. Culling of lions in small parts of their range is therefore of questionable value as an aid to increasing prey productivity in large conservation areas. Culling

[2] Prof. F.C. Eloff, Research Associate, Centre for Wildlife Management, University of Pretoria, South Africa 0002.

increases the proportion of cubs or subadults to adult lions, and these younger lions eat proportionally more on a body mass basis than do adult lions (Smuts 1978b). For culling of lions to be effective for increasing prey numbers in large conservation areas, the culling will have to be sustained for several years over most of the lion range (Smuts 1982). Furthermore, lions in the Kruger National Park can only regulate the numbers of their major prey when populations of the latter are low (Mills and Shenk 1992). In the southern Kalahari, all the predators present have an influence on their prey populations. However, the food supply available to the prey might be the eventual regulator of the prey populations in an area where the amount and distribution of rain has a great influence on the overall availability of food resources (Eloff 1984).

3.6.2 Leopards

Data on most of the population ecology of leopards in arid areas are virtually non-existent. Cub survival is low. Probably most adult leopards die of starvation. Cubs are also prone to starvation and predation by other predators (Bailey 1993). Infanticide of cubs has also now been confirmed in the Judean Desert where at least three different male leopards killed 11 cubs over a 9-year period (Ilani 1990; Bailey 1993). In the southern Kalahari, where a female has to fend for herself and her cubs, hunting for food may take the female away for up to 2 nights at a time. During this time, the cubs frequently die of starvation, thirst or are taken by other predators. Although solid data are still lacking, it is estimated that cub mortality may be as high as 90% there, especially in times of drought when prey is scarce (Bothma and Le Riche 1986). Young leopards are rarely seen in the southern Kalahari (Bothma, unpubl.).

Predation by other predators on leopard cubs is not rare. For example, in the Kruger National Park the remains of young leopards have been found in the stomach contents of spotted hyaenas (Bailey 1993). Several lions once treed a female leopard in the Serengeti National Park and devoured her two cubs (Schaller 1972). Adult leopards are occasionally killed by other large predators such as lions, spotted hyaenas, wild dogs, crocodiles and even by their own intended prey such as baboons *Papio* spp., warthogs or gemsbok (Eloff 1973a; Scott 1985; Bailey 1993). Porcupine quills lodging in the palate of a leopard when feeding may well lead to death (Bailey 1993).

In the Kruger National Park, the leopard population is probably self-regulated thanks to abundant food and a stable population of adults (Bailey 1993). In the wild, a female leopard might live to about 12 years of age (Scott 1985). When she dies, she is replaced by a female from the same area. However, when an adult male dies, he is replaced by a younger male from an adjacent area (Bailey 1993).

3.6.3
Cheetahs

Cheetahs have the highest juvenile mortality rates yet recorded for any cat that is not hunted by man (Caro 1994). It has already been shown that cheetah cubs have high mortality rates with less than 50% of all cubs surviving. This is especially due to predation by lions, which take 73% of all cheetah cubs between birth and independence (Laurenson 1995). In addition, some 50% of all the subadult male cheetahs in the Serengeti are killed by intraspecific fights when competing for territories. Usually, however, predation by lions, spotted hyaenas and leopards are only a secondary cause of cheetah mortality. Other causes such as disease, starvation or injuries are more important. To compensate for these high mortality rates, cheetahs reach maturity earlier, and have higher mean litter sizes than do other felids (Caro 1994).

The mean life expectancy of a female cheetah reaching three years of age in the Serengeti, is 3.9 years (Laurenson 1995). In the Nairobi National Park, there appears to be a relationship between the size of the cheetah population and the numbers of other large predators (Eaton 1970a). Cub recruitment of cheetahs in the Serengeti is currently insufficient to maintain the cheetah population found there (Laurenson 1995). However, increases in food resources may increase cub survival (Hanby and Bygott 1979).

3.6.4
Other Cats

In the arid Karoo region, the caracal selects rock hyraxes as a food resource in a density-dependent way. Although nomadic, the caracal is considered to be a main regulator of the rock hyrax and rodent numbers there (Palmer and Fairall 1988).

3.7
Interaction with Other Carnivores

Different carnivores usually interact to varying degrees, depending mainly upon their food requirements and the food resources available. In an ecosystem such as the Serengeti, the different sizes of the predator species present, their distribution over different habitat types, and the variety of their hunting or foraging methods all lead to a low degree of overlap in the food resources eaten. The extent of exploitation competition when carnivores compete for the same prey is therefore slight. Pursuit hunters, such as cheetahs, wild dogs and spotted hyaenas, take the slower individuals of their prey species, while ambushers such as lions take a more random selection. Interference competition, when carnivores compete for the same prey individual is usually more severe (Bertram 1979).

In adaptive carnivores, increasing prey scarcity may result in range expansion by the predators dependent upon them. For example, during one severe drought in the southern Kalahari, the available prey became widely dispersed and one lion pride once increased its range from 702 to 3900 km^2 in response to this (M.J. Owens and Owens 1984). Such conditions bring the predators involved into increasing contact with other predators into whose range they would not normally intrude.

Whether scavengers are a constraint upon predation by large felids, depends upon the abundance of these scavengers in a particular ecosystem, and upon the number of prey lost to them. Intraspecific competition for kills among solitary felids seems to be rare, and the possibility that they will meet at a kill is reduced by their temporal and spatial isolation (Sunquist and Sunquist 1989).

Of the felids in the arid bushveld of South Africa, the lion is the leopard's closest competitor, and there is a considerable degree of prey overlap (Kruger 1988). In the Etosha National Park, however, it is the spotted hyaena which seems to evoke most aggression from lions (Stander 1992b). In the Savuti Marsh, kleptoparasitism by spotted hyaenas of lion kills is a major competitive force. The prides there lose almost 20% of their food to spotted hyaenas, which live in clans of up to 50 members each. These hyaenas locate 78.5% of all lion kills, usually within 30 min of the kill having been made, and before much feeding has been achieved by the lions. On 82.4% of these occasions the hyaenas mob the lions at the kill, provided that the hyaenas outnumber the lions by at least 2:1. Complete take-over occurs regularly when they outnumber the lions by 4:1 (Fig. 3.4; Cooper 1991). Even in the Serengeti, lions may lose 10% of their food to spotted hyaenas if there are fewer than four lions on a kill (Caraco and Wolf 1975).

In the Kruger National Park, there is relatively little contact between lions and leopards, but given the opportunity, lions will take a kill away from a leopard. However, such a leopard may just escape with its prey up a tree (Bailey 1993). In the Masai Mara National Reserve, lions kill leopards irrespective of age or size. Even a group of olive baboons *Papio anubis* can chase a leopard from its prey (Scott 1985). Spotted hyaenas discover about 50% of all leopard kills, which they are easily able to take away if the leopard does not succeed in caching, or has already cached in a tree. A single spotted hyaena attracted to a leopard kill is occasionally killed by the leopard and is not normally able to appropriate the kill. However, two or more spotted hyaenas will usually take over a leopard's kill (Scott 1985; Bailey 1993). African wild dogs in the Kruger National Park occasionally pursue leopards with the obvious intent to kill them. The leopards there are also occasionally robbed of their kills by wild dogs (Pienaar 1969).

Leopards occasionally kill and eat an adult cheetah in the Kruger National Park, but contact between these two predators seems to be rare (Bailey 1993). However, leopards and black-backed jackals take a particular interest in each other in many areas (Scott 1985). In the Kruger National Park, black-backed jackals are never observed scavenging around leopard kills (Bailey 1993), possibly because of the danger of being preyed upon by leopards (Pienaar 1969) as in the southern Kalahari (Bothma and Le Riche 1984).

Black-backed jackals often follow foraging leopards around in the southern Kalahari, barking at them in a peculiar way as if notifying all and sundry about their whereabouts. The leopard will occasionally turn on them and kill them. Brown hyaenas also follow the tracks of foraging leopards at times in a possible attempt to scavenge food remains. A large, male leopard in the southern Kalahari, however, is able to defend its kill with varying success against up to two spotted hyaenas that may harass it simultaneously (Bothma and Le Riche 1984). Smaller carnivores such as the black-backed jackal, bat-eared fox, aardwolf and even on occasion the cape fox and genet *Genetta genetta* are preyed upon by leopards in the southern Kalahari (Bothma and Le Riche 1994a).

The competition between leopards and cheetahs for food in the dry riverbeds of the Kalahari Gemsbok National Park is reduced by the proportional selection of age cohorts of the springbok killed. Also, in the dunes away from the riverbeds, the leopards use a wider selection of prey than cheetahs. Furthermore, the cheetah hunts mainly by day and the leopard by night. This further reduces interference competition (Mills 1990). Bertram (1974) suggests that the leopards of the Serengeti are an unobtrusive, but integral part of the food chain.

The cheetahs of the Serengeti lose 12.7% of their kills to spotted hyaenas, and even more food to other large carnivores (Caro 1994). Competition between cheetahs and lions is not rare in the arid bushveld of South Africa, but these lions do not rob the cheetahs of their prey as they are known to do elsewhere. This is probably because the lions and cheetahs there are separated in time but not in space, as are cheetahs and leopards (Kruger 1988). In these bushveld areas, as they do elsewhere, cheetahs most often hunt during the day when other large predators are inactive (Schaller 1972; Eltringham 1979; Bertram 1979). This reduces competitive exploitation. In the southern Kalahari, most other predators, including the black-backed jackal and even some vulture species, rob cheetahs of their food. The main food source for the cheetahs there is young springbok lambs. This food resource is also used by at least eight other types of predator in the southern Kalahari, from martial eagles to lions (Mills 1990). The African wild cat is preyed upon by larger carnivores and birds of prey (Grobler et al. 1984).

The ethology of predation has been reviewed by Curio (1976), and the interactions between predators and prey in arid lands by Cloudsley-Thompson (1996).

CHAPTER 4

Family Hyaenidae

4.1 Taxonomy, Distribution, and Status

Although the modern hyaenids belong to a small family of the order Carnivora, the Hyaenidae was formerly a large family of nearly 100 species, with a wide distribution. Of the hyaenids, the four extant and some 15 fossil species are best known to science (Werdelin and Solounias 1991; Wozencraft 1993). The Hyaenidae have existed for about 25 million years (Werdelin and Solounias 1991) and are most closely related to the felids and viverrids (Delany and Happold 1979; Gould 1981), despite the fact that the spotted hyaena was originally considered a dog-like creature when first described as *Canis crocuta* by Erxleben in 1771. It was renamed *Crocuta crocuta* by Kaup in 1828. Similarly the striped hyaena was originally named *Canis hyaena* by Linnaeus in 1758, but was renamed *Hyaena hyaena* by Brünnich in 1771. The brown hyaena has always been considered to be a hyaenid, and it was first named *Hyaena brunnea* by Thunberg in 1820. In contrast, the aardwolf was originally thought to be a viverrid *Viverra cristata* by Sparrman in 1783. It was renamed *Proteles lalandii* by Geoffroy Saint-Hillaire in 1824 (Werdelin and Solounias 1991).

Fossil remains of hyaenids have been found in Africa, England and other countries of Europe, Asia and North America (Werdelin and Solounias 1991). The Pleistocene hyaena which lived in what is now Great Britian, may have hibernated in caves as its bones accumulated in great numbers in narrow underground tunnels. No comparable accumulations of hyaena bones, fossil or recent, have yet been recorded from any African site (Sutcliffe 1970). In South African fossil deposits, the hunting hyaena *Chasmaporthetes nitidula* and a large form of aardwolf *Proteles* sp. were present along with fossils of the spotted hyaena *Crocuta crocuta* and brown hyaena *Parahyaena brunnea* (Turner 1993).

The earliest known hyaenid fossil, the small, viverrid-like *Progenetta* sp. from the early European Miocene, shows that despite their current abundance, and for most extant species of hyaena also their exclusivity to Africa, the invasion of Africa by the hyaenids is a fairly recent event. Hyaenids reached their peak in Africa in the early Pleistocene, a period during which they lived together with the sabre-toothed cats (Mills 1990). However, the sabre-toothed cat *Smilodon fatalis* of the Pleistocene of North America avoided contact between its canines and bone during killing and feeding as its canines were less specialized for abrasion than those of modern carnivores, especially hyaenas. This suggests that sabre-toothed cats had a different method of killing behaviour and diet and possibly little competition with the hyaenids of their time (Anyonge 1996).

The four current species of the Hyaenidae are divided into two subfamilies. The subfamily Hyaeninae contains the spotted hyaena *Crocuta crocuta* of Africa, the brown hyaena *Parahyaena brunnea* of southern Africa which is also the rarest large carnivore in Africa, and the striped hyaena *Hyaena hyaena* of Africa, Asia, the Middle East and even southern Siberia and the Caucasus (Kruuk 1976; Mills 1977; Wozencraft 1993). The aardwolf *Proteles cristatus* is the only member of the subfamily Protelinae and occurs in suitably arid habitats in the Southwest Arid Zone and the Southern Savanna (Von Richter 1972; Wozencraft 1993). Here it lives in two discrete macro-populations, one in southern Africa, the other in East Africa from southern Tanzania to southeastern Egypt. These two macro-populations are divided by a mesic gap of some 1500 km (Richardson 1987a). The aardwolf is one of Africa's most specialized carnivores (Skinner and Van Aarde 1986). Although its exact origins are still uncertain (Mills 1990), its fossil record is known only from South Africa. Despite being placed in the separate carnivore family Protelidae by some earlier authors, there is today sufficient chromosomal evidence of its hyaenid relationship for it to be included in the Hyaenidae. However, it must have split from the other hyeanids early in the evolution of the family (Werdelin and Solounias 1991). It is also believed that the aardwolf's distribution across Africa is determined largely by the availability of those *Trinervitermes* spp. harvester termites which are active throughout most of the year, and which form the bulk of the aardwolf's food (Richardson 1987b).

The striped hyaena is known from fossil evidence from Africa only, but no evidence exists as to whether it is the most primitive of the three members of the subfamily Hyaeninae. Based upon the fossil record, its invasion of the areas outside Africa, where it also occurs today, may be quite recent (Werdelin and Solounias 1991). The origins of *Crocuta* are also uncertain, but it is known that *Crocuta* separated from *Hyaena* in the late Miocene. *Hyaena* and *Parahyaena* originated in the late Miocene from a common ancestor in southern Africa, making them some of the oldest of the extant Carnivora in the world (Mills 1990).

The conservation status of the hyaenids varies. The spotted hyaena is considered to be safe, but the striped hyaena *Hyaena hyaena barbara* of Morocco and western Libya is considered vulnerable in its entire range. The brown hyaena and the aardwolf also occur on one or another of the CITES appendices (Wozencraft 1993). The brown hyaena is better adapted than the spotted hyaena and the other large carnivores of southern Africa to live in arid regions where large herbivores are thinly and erratically distributed (Mills 1990).

Spotted hyaenas especially are still loathed by many people as a result of three myths which occurred in ancient texts. The first myth concerns the partially scavenging habit of spotted hyaenas, as many people originally believed that these hyaenas dug up graves in search of corpses. It was also believed that spotted hyaenas were hybrids and that they were excluded from Noah's Ark because the ark contained only pure-bred animals. Their origin was then explained as an unnatural union of a cat and a dog after the Great Flood. Lastly, the false hermaphroditic appearence of these hyaenas has led to the belief that they are abnormal and therefore abhorrent (Gould 1981). Many members of the Zulu people of South Africa still associate the spotted hyaena with witchcraft. In the early days of Zululand, there existed a dreaded fraternity known as "Inswelaboya" which literally means "those without hair". These people were reputed to be secret murderers and dealers in human body parts who rode about at night on monstrous spotted hyaenas. Even today the spotted hyaena is still in great demand for traditional rites and medicines in Zululand and elsewhere (Deane 1962).

4.2 Social Behaviour

The hyaenids show a diverse array of social behaviour and live singly or in groups of many members. The spotted hyaena is especially gregarious and has a well-defined set of social interactions, with different individuals playing different roles (Mills 1990). They are also the most numerous of all carnivores in the Serengeti ecosystem (Hofer and East 1995), and are the first carnivores for which social groupings of this magnitude, as commonly found among primates, has been described. All the behavioural adaptations commonly found in the spotted hyaena also occur among striped hyaenas, but not as prominently (Rieger 1979a). Among carnivores, only the brown hyaena has a social system similar to that of the spotted hyaena (Frank 1986a). Although they are solitary foragers, brown hyaenas have an advanced and flexible social system, with intricate relationships between individuals and with carcasses as important socializing focal points (Owens and Owens 1978; Mills 1990).

4.2.1 Group Formation

Spotted Hyaenas. A spotted hyaena clan revolves around its adult females (Mills 1990). The clans in the Serengeti National Park live in large, stable groups with a median size of 47 adults and subadults (Hofer and East 1993a, 1995). The term *clan* for such a group was coined by Kruuk (1966). The maximum size of a spotted hyaena clan in the Serengeti is smaller than in the Ngorongoro Crater, where clans of up to 100 individuals are found (Kruuk 1966).

However, the Ngorongoro clans are less stable than those of the Serengeti (Hofer and East 1995). Each clan defends a well-defined territory which includes a communal den. However, movements in search of food outside this terrritory are known in the Serengeti (see below; Kruuk 1966; Hofer and East 1993a, 1995). A clan regularly changes the location of its communal den within its territory (Hofer and East 1993a).

Relatedness is low in a spotted hyaena clan, and the dominance system varies depending upon the sizes of individual group members and their relatedness (Bertram 1979). There is no simple linear social hierarchy in the clan (Bertram 1979). However, intrasexual dominance hierarchies are found in males and in females. Female spotted hyaenas dominate males (Frank 1986a; Hofer and East 1995). In the Masai Mara area, one female and her descendants dominate all other clan members. These matrilineal rankings are stable over time because maternal rank is inherited. Cooperation among related females probably is the basis of the matrilineal system. Increasing female aggressiveness may in part be mediated by increasing levels of androgen in individual females, which in turn lead to greater dominance over other females (Frank 1986b).

Young, female spotted hyaenas in East Africa remain with their natal clan for life. In contrast, young males of 2 years or a little older disperse when they become sexually mature (Hofer and East 1995), as they also do in the Masai Mara area (Frank 1986b). Adult males become nomads which join other clans for brief periods, or immigrate and become the permanent resident males there (Frank 1986a; Hofer and East 1995). The females of a given clan therefore form various matrilines of related females, but the adult, resident males originate from outside the clan (Frank 1986a).

By commuting to find migratory game herds, the spotted hyaena clans of the Serengeti ecosystem can maintain clan sizes above those where they are solely dependent upon resident food resources. In the Serengeti, a spotted hyaena clan's size is not limited by the vital resources available in its territory (Hofer and East 1993a), as it is elsewhere in its range. Agonistic encounters between spotted hyaena clans in the Masai Mara area are rare, but not so in the Ngorongoro Crater (Frank 1986a).

The spotted hyaenas living in the arid bushveld areas of eastern South Africa follow a similar general pattern as elsewhere, but with some exceptions. They often do not forage in large groups, but do so when hunting large prey in the Timbavati Private Nature Reserve (Bearder 1977). In contrast, spotted hyaenas in the Kruger National Park normally move around in small groups when hunting (Henschel and Skinner 1990). The females of a typical spotted hyaena clan in the Kruger National Park are also closely related as they are in East Africa, while the males consist of three social classes: resident natal males, a group which is apparently not found in East Africa; peripheral, immigrant males and central immigrant males. However, the general spotted hyaena system of male dispersal and female philogatry remains. The females in a clan are philopatric and maintain amicable relationships with each other and their young (Henschel and Skinner 1987).

Social dominance assures the females priority at food, but the segregation of the clan into cooperative relatives and subordinate immigrants assures the optimal distribution of all resources (Henschel and Skinner 1987; Holekamp et al. 1997). Cooperation amongst the adult members of the clan is important for territory defence and for the procurement of large carcasses or prey. However, dominance relationships ultimately determine how the food will be divided up. In the Kruger National Park, linear dominance in a spotted hyaena clan occurs at all times, but it is most strongly expressed at food. This situation differs from that found in East Africa where there is a scramble for any food, but it is similar to that found in the spotted hyaenas living in the extreme arid central Namib Desert (Henschel and Skinner 1987), and may well be an adaptation to life in an arid and relatively prey-poor environment.

Male spotted hyaenas in the Kruger National Park that leave their natal clans, attempt to join a neighbouring clan. There they are at first treated as territorial intruders. However, eventually they may gain clan membership of a low status, with the lowest feeding priority and precarious social relationships with the other clan members. These males are known as peripheral male immigrants, but some which succeed in maintaining a close relationship with the females may eventually become central male immigrants, gaining unrivalled breeding status in the process (Henschel and Skinner 1987).

Spotted hyaena clans as large as 50 individuals are occasionally found hunting together, but not at a kill, in the dry Nossob riverbed of the Kalahari Gemsbok National Park (Eloff 1964). However, the clans in the Kalahari are usually small, ranging from three to 13 adults and subadults (Mills 1990). Unlike the situation in the Serengeti (Hofer and East 1993a), group size in Kalahari spotted hyaenas is at least partially influenced by the relative richness of the food resources in the territory, and also by disease and social behaviour (Mills 1990). This is in part also reflected in the fact that the smallest spotted hyaena clan size in the food-rich Ngorongoro Crater is 2.5 times that of the smallest clan found in the Kalahari. However, compared with brown hyaenas, Kalahari spotted hyaena clans fluctuate less. In contrast to East Africa, immigration by spotted hyaenas in the southern Kalahari is relatively rare, but more so in females than in males. Emigration is the most important determinant of group size (Mills 1990).

The Savuti Marsh harbours large spotted hyaena clans of up to 62 members each. However, these clans break up into small groups of up to four hyaenas to hunt or forage (Cooper 1990a,b,1991). Clan and territory size in Savuti are much larger than in South Africa, although the hyaena density is the same. The reason seems to be a fluctuating, and at times migrating, prey base, and the sparsity of water in the dry season in Savuti (Cooper 1990b). This situation confirms the general concept that territory and group size tend to be flexible in social predators, being influenced by the availability of critical resources such as water and food (Macdonald 1983). This is further confirmed by the small clan sizes of spotted hyaenas of the Etoṣha National Park (Gasaway et al. 1989, 1991) and by those of the central Namib Desert (Tilson and Henschel 1986).

Spotted hyaenas in the central Namib Desert (Fig. 4.1) live in small clans of up to eight individuals each. Only rarely there do spotted hyaenas move about in groups of more than three individuals, however, although they will congregate at a carcass. The Namib Desert clans are often separated widely in space, are largely exclusive, and movement of individuals between clans occurs infrequently. The occasional intruder is usually tolerated, but is nevertheless subjected to periodic harassment by one or more of the resident clan members. Although clan cohesion and cooperative hunting permits the spotted hyaena to exist in the central Namib Desert, the clans there must be flexible to adapt to wide fluctuations in the dispersion, abundance and availability of food and water. These resources are all mainly found along the Kuiseb River Canyon, which forms a linear oasis in terms of such resources. The above factors are therefore the primary determinants for range and clan size of spotted hyaenas of the Namib Desert (Tilson and Henschel 1986).

It is clear that where prey and water resources are abundant and reliable, as on the fertile grasslands of East Africa, spotted hyaenas can live in large, territorial clans. However, on the arid southern African savannas and deserts where prey is scarce and the water resources are an unstable commodity, clan size is smaller and more variable (Cooper 1990b).

Fig. 4.1. Sossusvlei in the central Namib Desert

Striped Hyaenas. Among striped hyaenas, food abundance and availability also play an important role in group formation, or the absence of it (Kruuk 1976). Competition plays little role in determining group size (Bouskila 1984). In East Africa, the striped hyaena is solitary and rarely associates with others of its kind. As a consequence, these hyaenas have developed some behavioural adaptations to living alone, and visual displays during interactions with other striped hyaenas are infrequent (Kruuk 1976). In contrast, however, the striped hyaenas living on the coastal plains of the Dead Sea are seen in pairs on more than 50% of all occasions. On these coastal plains, striped hyaenas have only one main and clumped food source, which is visited by several individuals from the same or different social units (Macdonald 1978). The striped hyaenas of the Negev Desert commonly form groups of two to four individuals, often consisting of a mother and her young, or a group of litter mates moving around together on their own. One principal reason for group formation in the striped hyaenas of the Negev Desert, compared with those of East Africa, seems to be the operative pressures of the larger mean size of the main food items available in the Negev, and the greater quantity and availability of food for striped hyaenas in East Africa. In the Negev, with its poor food resources, striped hyaenas form foraging groups based upon shared information. When one hyaena finds food there, it immediately returns to the other members of its social group to lead them to the food (Bouskila 1984).

Brown Hyaenas. Brown hyaenas have been studied in depth, and particularly so in southern Africa (e.g. Mills 1990). Although Eaton (1976a) believes that these hyaenas are not truly social, most brown hyaenas in the Kalahari live in small, mixed-sex groups of related individuals (Owens and Owens 1979a; Mills 1982a,b). Numbers fluctuate markedly within and between clans (Mills 1990), which have a distinct social hierarchy (Owens and Owens 1979a,b). Nevertheless, little evidence has been found for a social hierarchy in the brown hyaenas of the southern Kalahari (Mills 1983). As in other hyaenas, group size in the brown hyaena there is affected by the quality, distribution and abundance of its food (Owens and Owens 1979a; Mills 1982c, 1983). However, as a major scavenger, it is not the biomass of potential food available in a given area which can be killed by other predators, hence making it available as carcass remains for brown hyaenas, which is important to group size (Mills 1982c, 1983). Most of the brown hyaenas of the Kalahari live in groups of varying size in well-defined territories which overlap little with those of the neighbours (Mills 1982c). Communal dens serve to develop the social ties in the clan, and the evolution in brown hyaenas of communal denning may itself have been stimulated by the utilization of a food resource whose exploitation demands much time and energy away from the den (Owens and Owens 1979a).

The most important group size variable for brown hyaenas in the southern Kalahari is subadult emigration (Mills 1983). Some subadults of both sexes leave their natal groups (Mills 1982b). The brown hyaenas there mostly live in small groups. Apart from isolated incidents of "dominance" at food and some other isolated cases, there is little evidence of a dominance hierarchy in brown hyaena groups in the southern Kalahari (Mills 1990). Group cohesion is maintained through other behavioural patterns, such as an elaborate greeting ceremony, allo-grooming and possibly some forms of muzzle-wrestling. Muzzle-wrestling also seems to be particularly important in regulating group size through subadult recruitment and expulsion. Aggression is most marked between individuals of the same sex from neighbouring territories (Mills 1983). In the central Kalahari, however, the brown hyaena is more social than previously believed (Owens and Owens 1979a). Brown hyaena group members there congregate in groups of up to six individuals at large carcasses for communal scavenging. They also use common pathways, maintained through scent-marking, to move around (Owens and Owens 1979b).

Aardwolves. The aardwolf is a highly solitary animal which is intolerant of the company of other aardwolves except when mating or accompanying young cubs (Richardson 1991). Aardwolves of the central Namib Desert also den and forage singly, only forming small groups when a female moves around with her cubs (Nel and Bothma 1983).

4.2.2 Nomads

Male spotted hyaenas that disperse from their natal clans usually become transients or nomads which attach themselves temporarily, or occasionally on a permanent basis, to other clans (Hofer and East 1995). In the Ngorongoro Crater, however, the spotted hyaena clans are highly territorial, and nomads are not as tolerated there as they are in the Masai Mara National Reserve (Frank 1986a). In all cases, nomads attaching themselves to other clans are at first subordinate (Frank 1986a), but a male can later become a central, immigrant male with an unrivalled breeding status (Henschel and Skinner 1987).

In the Serengeti, all large predators, including the lion, leopard, spotted hyaena, cheetah and African wild dog, are nomadic and follow the migrating game (Hofer and East 1995). However, this seems to be nomadism in a different context to that described above. In the brown hyaena population of the southern Kalahari, up to 8% of the individuals are nomadic, breeding, adult males which later join other clans. Up to 30% are dispersing subadult males (Mills 1982c). As in spotted hyaenas, these nomadic males have a disregard for territorial boundaries, and have no lasting relationship with conspecifics. However, unlike in spotted hyaenas, nomadic, brown hyaena males may be reproductively active. They are important in the mating system as they mate most with the group-living females (Mills 1983, 1989, 1990), and are therefore not on the low end of the social scale as is the case in spotted hyaenas. As in spotted hyaenas, however, an immigrant male will eventually become accepted into a clan if it persists in returning, hanging around the den and coming in on kills. However, this process may take several months to complete.

Once assimilated into the clan, immigrant males play an important role in foraging, and they frequently lead the foraging group (Mills 1990). In the Savuti Marsh, nomadic, dispersing, subadult spotted hyaena males, older and non-breeding males, and possibly a few females become nomads or transients that hunt in foreign territories, but remain subservient to the residents of such an area (Cooper 1990b). In the central Namib Desert, the presence of temporarily nomadic, solitary spotted hyaena males on the northern gravel plains, suggests that some spotted hyaenas, especially males, may survive in marginal habitats where ecological conditions do not permit the existence of clans (Tilson and Henschel 1986).

4.2.3
Communication

Hyaenas use a variety of tactile, vocal and olfactory systems of communication. In the context of olfactory communication, they are known to use latrines, to paste anal gland secretions onto objects in their range, to make scrapes that are scented with interdigital gland secretions, and to roll or rub on decayed carcasses. Pasting is unique to hyaenids. It is the secretion of a strong-smelling substance from the anal glands, usually onto grass stalks (Mills 1990). The term *pasting* was coined by Kruuk (1972b). Urine does not seem to have a scent-marking function in hyaenids (Owens and Owens 1978). Scent-marking by hyaenids, as by other carnivores, has more than one function in any one species, and different functions in different species (Bearder and Randall 1978; Mills1990).

Spotted Hyaenas. The whooping call of the spotted hyaena is one of the call-signs of Africa. Vocal signals or calls are used by a variety of mammals to maintain intergroup spacing and terrritory integrity. In spotted hyaenas, whoops also serve to identify individuals. Unlike other social carnivores, such as lions that frequently vocalize as a group, spotted hyaenas mostly call individually, and usually do so in self-advertisement (East and Hofer 1991). It is a method of terrritorial defence, but is also used for other aspects of communication (Hofer and East 1995). Spotted hyaenas in the Serengeti National Park use a loud, whooping call which has three main functions. It displays individual identity, requests support, and conveys information about the location of the caller. Whoops differ according to the sex and age of the caller, and the reason for calling. Female spotted hyaenas whoop during agonistic encounters with other female members of the same clan, when clashing with neighbouring clans, when trying to locate their young, and when rallying clan members to defend a communal resource. Breeding females also whoop to deter potentially infanticide-minded non-clan members from approaching the cubs at the den. Adult males whoop more often than females do to display their identity, especially during inter- and intrasexual display. Cubs primarily whoop to provide information about their location and to request help (East and Hofer 1991). Young cubs also have a call that is almost identical to the chuckle of the adults, but which is sharper and more high-pitched (Pournelle 1965).

In the arid bushveld of the Timbavati Private Nature Reserve, spotted hyaenas use calls to identify food sources with pin-point accuracy (Bearder 1977). By rolling on decayed carcasses, they also communicate the presence of food to their companions during normal meeting rituals (Bearder and Randall 1978). In the Kruger National Park, whooping is seldom used by spotted hyaenas to advertise their territory. However, it is used to advertise space use rather than as a mechanism of space protection or defence (Henschel and Skinner 1991). Scent-marking is the main method of territory defence there (Henschel and Skinner 1987).

Spotted hyaenas of the southern Kalahari use vocal signals as one method of conveying their mood and status to conspecifics. Vocal calls are used to signal specific events, or can be spontaneous (Mills 1990). However, in addition to various visual displays, these vocal signals are important in maintaining a complex social organization, especially as some individuals may change rapidly from being solitary to highly social. Other hyaenids do not possess this wide range and frequency of behavioural patterns (Mills 1990).

It is believed that changes in vocal display with dominance in spotted hyaenas may be limited when the clans are small and dominance hierarchies involve a few individuals only. This happens in some parts of the arid Kalahari ecosystem. Most of these clans have only one adult male. This eliminates competition between males, and in turn should lead to fewer displays and less display effort on the part of the males, when compared with the situation on the East African grasslands. In the Kalahari, physical conflict between males over females in oestrus should also be rare. Possibly the vocal displays of male spotted hyaenas there only serve to discourage nomadic males from joining a clan (East and Hofer 1991).

Urine does not appear to have any function of communication in spotted hyaenas (Owens and Owens 1978). However, accumulations of spotted hyaena faeces form latrines which have a definite communication function (Mills and Gorman 1987). Latrines in the southern Kalahari are clumped near to, but not along the boundary of the territory (Gorman and Mills 1984). However, not all the faeces are deposited exclusively in these latrines (Deane 1962). In the Kalahari Gemsbok National Park, latrines tend to be concentrated along the dry riverbeds which form a major habitat for the spotted hyaenas of that region. It is also known that the spotted hyaenas of the southern Kalahari scent-mark more often in those areas most frequented by them, and where most food occurs. Their large terrritory size there makes effective border marking of a territory impossible to achieve (Mills and Gorman 1987). In the Kruger National Park, latrines are one of the terrritory demarcation methods of spotted hyaenas (Henschel and Skinner 1987). In the central Namib Desert, spotted hyaenas use latrines as one method to scent-mark the core-areas of their range. In these core-areas, the breeding dens, lairs and water resources are located. The latrines are not as small as those of the brown hyaenas of the Namib, and may cover an area of some 100 m^2. Some latrines are used by the whole clan, and others by certain individuals only (Skinner and Van Aarde 1981). In addition, the main trails which fan out from the core-areas of the range are scent-marked with latrines, and other methods such as pasting. In one clan in the Namib, 91% of all scats were found within a 200 m radius of the den, and 72% in another (Tilson and Henschel 1986).

The highly territorial spotted hyaenas of the Kruger National Park spend 20% of their total activity on territory patrol, during which they scent-mark intensively and monitor such marking points. One clan there is known to have 64 marking points in its territory, most of which occur on the boundaries of the territory (Henschel and Skinner 1991). These scent marks involve scrapes, pasting and often latrines (Henschel and Skinner 1987). Patrolling hyaenas in this park keep to game trails, dry riverbeds and roads more often than foraging hyaenas do, and scent-mark 16 times as often when they patrol as when they forage (Henschel and Skinner 1991). Spotted hyaenas in the Ngorongoro Crater only scent-mark their territory boundary. This is done because the territories there are relatively small. The boundaries are therefore more easy to patrol and mark (Mills and Gorman 1987).

Both spotted and brown hyaenas in the southern Kalahari scent-mark their territories by pasting excretions from their subcaudal scent glands on to grass stems, and by defecating at latrines. However, they adopt different strategies for dispersing these scent-marks in their territories. In the spotted hyaena there, 62% of all the pastings are placed at latrines, while others are placed on some conspicuous environmental feature elsewhere in the whole territory. In this arid environment, the scent-marking pattern of the spotted hyaena is an adaptational response to the problem of marking a large territory with a limited amount of scent and within a limited time (Mills and Gorman 1987).

In some hyaenids there is no dominance structure, and no difference between social classes in pasting frequency (Mills 1990). In spotted hyaenas in the southern Kalahari, adult males paste significantly more often than an adult females or subadult males. However, subadult females and adult males paste at the same rate (Mills and Gorman 1987). Of all social classes, however, it is the adult, immigrant males that paste most frequently despite the fact that they are subordinate to the dominant females. This is probably because such males must advertise vigorously to be able to join a clan (Mills 1990). The pasting rate increases considerably in adult male and female spotted hyaenas that move long distances to the periphery of the territory (Mills and Gorman 1987). Spotted hyaenas paste at a mean rate of once every 7.7 km, but when foraging, they paste more frequently. An individual spotted hyaena in the southern Kalahari will usually be 1 to 2 km away from a scent-mark over most of its territory, and within 500 m of one when near the centre of it (Mills 1990). The scent-marking strategies of spotted hyaenas in the Kalahari, and in the arid bushveld of eastern South Africa, are quite different from those in the Ngorongoro Crater where environmental conditions are also quite different. Scent-marking in spotted hyaenas is therefore not species-specific, but is determined by the ecological pressures which act on different populations (Mills and Gorman 1987), and of which aridity is one.

The fact that spotted hyaenas usually forage and feed together has influenced their behavioural patterns, and affected the system of communication. Spotted hyaenas may be able to distinguish between the pastes of different individuals, and they react intensively to alien pastes. This is consistent with the concept that scent-marks function to allow the unambiguous recognition of those individuals that invest most heavily into a given area (Mills 1990). Spotted hyaenas only have one type of paste as opposed to the two found in brown hyaenas and aardwolves (Nel and Bothma 1983; Mills 1990).

The spotted hyaenas living in the Savuti Marsh also scent-mark mainly on their territory boundaries, but latrines are uncommon there (Cooper 1990b). Pasting and scrapes often occur at latrines and are used extensively by spotted hyaenas in the central Namib Desert. They mark the core-areas of their range and the trails which lead away from these core-areas (Bearder and Randall 1978; Tilson and Henschel 1986). Intergroup encounters, and associated territorial behaviours as seen among East African spotted hyaenas, are not found in the Namib Desert (Tilson and Henschel 1986).

Striped Hyaenas. Little is known about communication in the striped hyaena. A striped hyaena does have a range of vocalizations, however, which resembles that of the spotted hyaena, but a spotted hyaena's vocalizations are louder, more varied and more frequent in comparable situations (Kruuk 1976). A striped hyaena pastes exactly as a spotted hyaena does. The brown hyaena appears to be intermediate between the spotted and the striped hyaena in terms of overall scent-marking. A striped hyaena uses faeces (latrines) to mark certain areas of its range, and pastes anywhere in its range, but usually not close to its den as a brown hyaena does (Kruuk 1976; Owens and Owens 1979a). Like the spotted and brown hyaena, urine does not seem to function in communication in the striped hyaena (Kruuk 1976).

Brown Hyaenas. Brown hyaenas usually forage alone. This has influenced the evolution of many of their behavioural patterns as they tend to interact with conspecifics singly. All the brown hyaena's vocalizations involve short-distance communication, although they can be heard over several kilometres. The hoot-laugh is the only discrete call (Mills 1990). Brown hyaenas in the central Kalahari recognize each other individually. In a clan there is a well-defined ranking order which is maintained through neck-biting and other social interactions which include scent-marking, vocalization, visual displays and elaborate rituals. Individuals meet regularly and do not avoid each other (Owens and Owens 1978).

The latrines of brown hyaenas in the southern Kalahari also have a scent-marking function. They occur over the whole territory, but are concentrated near the territory boundaries, and next to prominent landmarks such as the shepherd's tree. Brown hyaenas visit those latrines significantly more often that are located near the territory boundary. Many latrines, however, are only used for short periods (Mills 1990). The life-span of a latrine is unknown, and may possibly vary (Gorman and Mills 1984). In the central Namib Desert, the latrine of a brown hyaena is small and clearly demarcated. It is probably only used by one individual and not by a whole clan as in the case of the spotted hyaena (Skinner and Van Aarde 1981).

Although Gorman and Mills (1984) reported that the brown hyaena is unique among hyaenids in producing both a long-acting and a short-acting secretion from the anal glands, the same phenomenon had previously been found in the aardwolf (Nel and Bothma 1983). At that time the aardwolf was already considered by most zoologists to be a member of the Hyaenidae (Smithers 1983). The two types of secretion, and the higher pasting frequency of the brown hyaena as opposed to the spotted hyaena, suggest a broader function for pasting behaviour in the brown hyaena. The short-acting, watery, black secretion may have an intergroup function. As brown hyaenas often feed on small items with a low renewal rate, it is important that every member of a clan should know where the other members of the clan have recently foraged so that unproductive foraging efforts can be avoided (Mills and Gorman 1987). The short-acting secretion seems to be individually unique and recognizable (Mills 1990).

Long-acting pastes of the brown hyaena are found throughout a territory in the southern Kalahari, but more often in the central areas where the hyaenas spend most of their time (Gorman and Mills 1984). The large size of the territory of the southern Kalahari brown hyaena makes effective scent-marking of its boundaries just as impossible as it is for spotted hyaenas there (Mills and Gorman 1987). Paste-marking by brown hyaenas is a significant investment of energy. The white, long-lasting paste is 97% lipid and an individual brown hyaena may make as many as 2900 such pastes in a year (Gorman and Mills 1984). They mark 2.6 times in every kilometre, or once every 384 m when travelling. This is 20 times more often than the scent-marking frequency of a spottted hyaena there. A brown hyaena in the southern Kalahari is at most 250 m away from a scent-mark over most of its territory, and within 500 m of one over the entire terrritory (Mills 1990). The length of the boundary that must be marked by each individual in a brown hyaena group is an important factor. It determines whether the whole territory will be marked, or if it will only, or mainly, be marked along its boundary (Gorman and Mills 1984).

A brown hyaena can probably distinguish between the pastes of the individuals of its own group, as well as that of members of a different clan (Mills 1990). The number of active scent-marks in a territory at any given time depends on the ratio between the rates at which new scent-marks are laid down and old ones decay. Pastes are arrayed vertically on objects at the normal nose height of a hyaena to maximize their chances of discovery (Gorman and Mills 1984).

As the brown hyaenas of the central Kalahari use common pathways maintained through scent-marking, they often meet other brown hyaenas when foraging (Owens and Owens 1979b). Latrines are usually located under dense bushes, but are not as abundant as those of spotted hyaenas. Pasting, however, is the most important method of communication, and takes place every 1 to 3 min during foraging (Owens and Owens 1978).

Aardwolves. Scent-marking is well developed in the aardwolf. This carnivore has two types of anal gland secretion. It defecates in clearly defined latrines or middens where the faeces are covered with sand. Middens may possibly have no scent-marking function, although the aardwolf well knows where these middens are located (Nel and Bothma 1983).

In the arid Northern Cape Province, the aardwolf uses scent marks to advertise its territory. These scent-marks are concentrated along territorial boundaries and occur in increasing densities the smaller the territory is. They provide olfactory information for the resident aardwolf, and on its defended area for others. The message relayed depends heavily on the size of the territory, the frequency of territorial encounters and the defensive behaviour of the resident. The greater the territory size, and hence the greater the difficulty in monitoring intrusions, the greater the threat that a message will transmit. It may even lead to a physical attack on an intruder (Richardson 1991).

It has often been proposed that scent marks in mammals are characteristic of the individual producing them. Hence they function as an unambiguous signal of territory ownership and establish an important imbalance in the contest between a resident and an intruder (Gosling 1982). In aardwolves, a territorial resident will mark over the scent marks of an intruder. Aardwolf scent-marking helps to maintain, but cannot ensure, territorial integrity during the temporary absence of the territorial resident. However, in the presence of the resident, the scent marks are fresh. Indirect threats of aggression then change to direct ones. Therefore a change in perception occurs, which is of survival value to any intruder (Richardson 1991).

During the breeding season an aardwolf retains its territory. However, there are also dramatic changes in the social and scent-marking behaviour. These changes are related to the critical resources available. They also complicate the pattern of territory maintenance, a process which is usually stable when outside the mating season (Richardson 1991).

Aardwolves usually scent-mark on grass stalks. Adults scent-mark little outside their territories, doing so in only 5% of all pastes in the Northern Cape Province. However, some aardwolves do make excursions outside their territories, but when they do so, they stop scent-marking. Boundaries are marked significantly more often by both sexes than other parts of the territory. The male aardwolf spends more time at the territory boundaries than elsewhere, while the female spends more time at the centre (Richardson 1990).

The anal gland secretion and pastes of the aardwolf consist mainly of short, medium-chain, fatty acids whose components occur in markedly variable concentrations. The odour of a fresh paste has a dual potential. It can act as a short-term indicator, while the long-term persistence of odour may enhance its effectiveness in demarcating territory (Apps et al. 1989). Aardwolves in the central Namib Desert produce two types of anal gland secretion for short-acting and long-acting scent messages respectively. The short-acting paste is small and easily missed by the human eye. It is used mainly for the recognition of foraging areas already used in a given night, and are left frequently at 5 m intervals over the foraging range. The long-acting paste is more commonly found at faecal middens and dens, and it serves to identify important areas of the territory. Moreover, the marking rates of any given individual vary between nights, as they do between different individuals on the same night (Nel and Bothma 1983).

4.3 Mimicry

Mimicry is a condition in some species and sexes of animal in which one has the appearance of another. In the hyaenids there are two forms of mimicry. They are physical and sexual mimicry, the latter being the more known.

4.3.1 Physical Mimicry

The aardwolf is not totally sympatric with the striped hyaena, yet the two have a striking morphological similarity. Some authors regard this similarity as mimicry, especially where the two species occur sympatrically. In this context, it is interesting that the aardwolf of southern Africa, where the striped hyaena does not occur, looks distinctly different from the one in the north, where striped hyaenas do occur, although no subspecific divisions are recognized (Eaton 1976b).

Of all the sympatric carnivores living in southern Africa, the aardwolf most closely resembles the brown hyaena. However, this resemblance is not as close as it is further north. In southern Africa, the aardwolf may mimic the brown hyaena. As the brown hyaena occasionally dominates some possible predators of the aardwolf, such as the caracal and cheetah, this mimicry may benefit the aardwolf by increasing the caution with which such predators will approach it. For mimicry to have evolved, complete immunity from attack is not required, but only an increased probability of survival (Eaton 1976b). This subject has been reviewed extensively by Cloudsley-Thompson (1996).

4.3.2 Sexual Mimicry

The phenomenon of sexual mimicry is especially well-developed in the spotted hyaena. It has long been known that there is something peculiar about the sexual anatomy and reproductive physiology of the spotted hyaena (Matthews 1939). There is a belief which was held by many ancient writers, but not by Aristotle (384 to 322 B.C.), that the spotted hyaena is a hermaphrodite (Matthews 1939; Gould 1981). However, Aristotle did not distinguish between the spotted and the striped hyaena, and the legend only relates to the former (see below). Pliny the Elder (A.D. 23 to 79) repeated the legend, but mentioned Aristotle's disagreement. Claudius Aelianus (A.D. 160 to 220) also stated that a spotted hyaena would change its sex in alternate years. Even as recently as 1936, Hemmingway referred to the spotted hyaena as a hermaphrodite (Matthews 1939).

The above myth of hermaphrodism only applies to the spotted hyaena. A female striped hyaena is clearly recognizable as such (Wells 1968), and the ancient writers did not know of the brown hyaena and aardwolf. They therefore could not refer to them. The fallacy of hermaphrodism is based upon the appearance of the sexual organs of male and female spotted hyaenas which are strikingly similar externally. Female spotted hyaenas have a peniform and highly erectile clitoris, and a false scrotum (Matthews 1939; Deane 1962; Racey and Skinner 1979; Van Jaarsveld and Skinner 1987). In sexual appearance, a male spotted hyaena is externally indistinguishable from a non-parous female. Moreover, the female is larger than the male (Matthews 1939). The larger size of the nipples of a parous female is the sole characteristic by which to distinguish her sex without dissection (Matthews 1939).

Detailed studies now suggest that high foetal androgen levels are responsible for the male appearance of the sexual organs of the female spotted hyaena (Racey and Skinner 1979). They lead to over-development of the sexual organs of the female through their influence on embryonic development (Gould 1981; Frank et al. 1995). The high plasma androgen level found in female spotted hyaenas may also be associated with their overall aggression and dominance over males. Furthermore, the appearance of the female's sexual organs is probably also related to behavioural sex, as it is in male cercopithecoids and female ceboids (Racey and Skinner 1979). Genetic translocation from the Y- to the X-chromosome would also expose the female foetus to raised testosterone levels, leading to social dominance by the adult female (Van Jaarsveld and Skinner 1987). These factors all lead to a system whereby the female dominates and leads the clan, gaining maximum benefit from her size and aggression (Gould 1981). In an ancestral spotted hyaena which was busy making the behavioural transition from *Hyaena*-like scavenging to a life of active predation, there would have been an increase in feeding competition. Large, dominant and aggressive females which were able to out-compete other hyaenas for food, would have been more successful in rearing cubs as a result of their elevated nutritional status, and the competitive advantage which they conferred to their young (Frank 1986b).

4.4 Reproduction

4.4.1 Spotted Hyaenas

There is general consensus that the spotted hyaena is polyoestrus, with no clearly defined breeding season. Breeding varies locally, with minor peaks which are possibly linked to local patterns of food abundance (e.g.: Matthews 1939; Deane 1962; Racey and Skinner 1979; Lindeque and Skinner 1982; Frank 1986a; Cooper 1993). The complete sexual cycle lasts nearly 12 months, with pregancy and lactation together persisting for 9.5 months, and anoestrus for 3 weeks to 2 months (Matthews 1939). Gestation lasts for 110 days (Van Jaarsveld et al. 1988; Frank et al. 1995). In the absence of fertilization, oestrus occurs again after 14 days (Matthews 1939). The female begins reproducing when about 3 years old, and has a reproductive life-span of 13 years. The male reaches reproductive maturity when 2 years old (Frank 1986a; Frank et al. 1995; Holekamp et al. 1997). The mating system is highly polygenous and only the dominant male of a clan mates, although all resident males regularly court females. Among the females, there is no correlation between reproductive success, and age, size or social rank (Frank 1986b). In the Masai Mara area the mean inter-litter interval is 14.5 months (Frank 1986a).

Newborn spotted hyaena cubs are dark brown and weigh about 1.5 kg (Pournelle 1965). The sex ratio at birth in the southern Kalahari favours females (0.4 males per female), and litter sizes are small (mean 1.7) (Mills 1990). The eyes are open at birth. Eruption of the molars starts at 31 days of age and replacement of the juvenile pelage when 7 weeks old. The moult may only be complete at 9 months of age. The cubs grow rapidly and can double their birth weight in 25 days, tripling it in 37 days. When 100 days old, a spotted hyaena cub weighs nine times what it did at birth (Pournelle 1965). Postnatally, however, a spotted hyaena grows slowly and only reaches its asymptotic weight when about 10 years old (Van Jaarsveld et al. 1988). The cubs are suckled for 14 to 18 months (East and Hofer 1991), a period which places a severe lactating strain on the females (Hofer and East 1993b,c). The reproductive success of the dominant, or alpha female is measured in terms of the number of male young that she produces. The sons of the alpha female show behavioural patterns which differ from those of other males in that they are less submissive to dominant females when they become adult, and that they leave the natal clan at an older age than the other males. However, same-sex litters generally lead to severe fighting among litter mates, often resulting in the death of one of the siblings (Werdelin and Solounias 1991).

Spotted hyaena cubs are born in a natal den (Mills 1990). For the first 2 weeks of life they are poorly coordinated and rarely venture outside the den (East et al. 1989). The denning and suckling periods of hyaenas are longer than in any other large carnivore. For the first 12 to 15 months of their lives, hyaena cubs remain at the den (Mills 1990). The use of a natal den offers some form of protection for the cubs, but also entails the risk of death when the cubs have to be transferred to the communal den of the clan. This takes place when the cubs are about 2 weeks old. The typical natal den consists of an entrance cavity, which tapers into a narrow tunnel approximately 200 mm in diameter. A female crouching in the entrance cavity is invisible from the surface and is sheltered from the sun. However, only the cubs can pass through the narrow channel to the deeper and wider portion of the den. This design allows the cubs to evade large, threatening predators such as lions. Communal dens have a similar design, but several den cavities are clustered in one area. Judging by the visits of other clan members to natal dens, these dens are centres of social activity as are the communal dens (East et al. 1989).

When younger than 4 weeks old, spotted hyaena cubs have to be moved between dens. They are generally carried by their mother. However, in the Serengeti ecosystem a young cub approximately 6 weeks old once walked 14.6 km in 4 h 20 min, following its mother towards a new den. The mother kept the cub within 3 m of herself and carried it occasionally to help it keep up. This was to no avail, however, as the cub was killed by a lion before reaching the new den (East et al. 1989).

Except for the brief stay in the natal den, a spotted hyaena cub in the Serengeti spends the first year of its life at the clan's communal den where it remains dependent on suckling to survive (Hofer and East 1993b). Several litters will be raised in such a communal den (Mills 1978a). Cubs in the Serengeti leave the communal den at a median age of 357 days (range: 244 to 481 days). At the communal den, a mother will visit her cubs almost every morning and evening. She actively defends her cubs till they are about 4 weeks old, but thereafter the frequency of her visits declines. Especially when the mother has to commute great distances to a remote food source, the cubs continually lose weight while the mother is absent. When she returns they imbibe great quantities of milk to compensate for mass losses during her absence (Hofer and East 1993b). Spotted hyaena cubs are weaned when they are 12 to 16 months old (Mills 1990).

The mothers of single cubs in the Serengeti attend their cubs with the same frequency as do the mothers of twins, but the growth rate of a single cub is higher than that of each of twins. As they grow faster, single cubs also have higher long-term survival rates than twins. The available prey resource in a clan's territory influences the pattern of cub growth. This depends upon whether there is a need for the mother to commute long distances over several days to the nearest food. Mean seasonal growth rates of whole litters also decline as the mean absence interval of females at the den increases, but increase with increasing female attendance at the den. As the cubs grow older, the porportion of visits by denning females decreases, while the period of absence increases (Hofer and East 1993b). Adult spotted hyaenas do not carry food to the den to feed young cubs, but they sometimes do so for their own consumption. Then the cubs may opportunistically share in the food (Mills and Mills 1977; Hill 1980; Skinner et al. 1986; Mills 1989). When older, the cubs of dominant or high-ranking females are able to feed at kills more successfully in competition with adults than other cubs (Frank 1986b).

Spotted hyaena females in the Kruger National Park have litters every 16 months, and may lactate for as much as 75% of their lives. As cubs are solely dependent on milk for at least the first year of life, lactating females have to ensure a high protein intake to keep up with this great demand for energy (Henschel and Skinner 1987). Periods of food abundance for a spotted hyaena usually coincide with the weaning of the cubs, thus allowing the females to conceive again when in good physical condition. However, the ability of females to respond to seasonal periods of food abundance can never lead to permanent adaptations in reproductive strategy because the normal female cycle of reproduction exceeds a year. Therefore, every year a different group of hyaenas takes advantage of seasonal food abundance, such as the annual impala lamb crop (Lindeque and Skinner 1982).

Cooperation in raising young in the southern Kalahari is less apparent in spotted than in the brown hyaenas. Spotted hyaena dens there have more entrance tunnels than those of brown hyaenas because they contain more cubs. Furthermore, spotted hyaena dens do not have auxiliary dens close by as is the case with brown hyaenas. Den occupancy by spotted hyaenas is relatively short. This contrasts sharply with the situation in the Kruger National Park where spotted hyaenas spend six months or more, and even sometimes several years, in the same den (Mills 1990). Once subadult, a male has three main reproductive options: to remain with the natal clan, to become nomadic, or to immigrate. A female that becomes subadult also has three reproductive options: to remain with the natal clan, to immigrate, or to establish a new clan (Mills 1990).

As elsewhere, spotted hyaenas in the southern Kalahari do not carry significant amounts of food to the den. What they do bring is only skin or bones. The spotted hyaena cubs therefore rely on their mother for milk until they start foraging with their mother when a year old. Lactating, female spotted hyaenas are said only to suckle their own cubs (Cooper 1993), but in the southern Kalahari some evidence has been found of allo-suckling, with different females suckling the same cub. Nevertheless, this remains a rare event (Mills 1990). It is just possible that the aridity of this environment with its prey-poor resources, and the large range sizes of spotted hyaenas there, could be responsible for occasional allo-suckling in brown hyaena cubs as a survival mechanism.

The cubs of spotted hyaenas in the Savuti Marsh also use a night den in addition to the usual communal den. Such a night den is usually located near a main hunting area. Once a cub is three months old, and therefore mobile, it is brought from the communal den each evening to the night den, usually a disused warthog burrow. There it stays in the presence of a few adults or subadults while the rest of the clan hunts. At dawn, the mothers lead their cubs back to the communal den. No active defence of a den containing cubs is known to occur. Cubs which are threatened, simply move underground and out of danger (Cooper 1993).

Although the cubs of a spotted hyaena are relatively safe while close to their den, they do suffer mortality from time to time. However, despite the harsh environment of the Kalahari, cub mortality there is low (Mills 1990). Same-sex litter mates sometimes fight severely from birth, resulting in the death of one of the siblings. Such siblicide, which is common in large raptors, is unknown in other mammals and its selective function is unknown (Werdelin and Solounias 1991). No incidence of infanticide by adult spotted hyaenas has been witnessed (East et al. 1989). There also seems to be no tendency for females of different social rank to experience different rates of litter loss before the cubs are six months old (Frank et al. 1995).

4.4.2 Striped Hyaenas

Reproductive activity in striped hyaenas begins at two to three years of age when sexual maturity is reached (Rieger 1979a). As in the brown hyaena, the striped hyaena is polyoestrus and breeding occurs throughout the year (Skinner and Ilani 1979). The gestation period is 88 to 92 days and the mean litter size is 2.4 (range: 1 to 5). Striped hyaena cubs open their eyes when five to nine days old and first emerge from the natal den when they are 10 to 14 days old (Rieger 1979a). The natal den is usually located in a small cave, or in a cavity in a rocky area, if available. The young cubs are visited frequently by their mother and suckle till they are 10 to 12 months old. As a brown hyaena does, but unlike a spotted hyaena, a striped hyaena female regularly carries meat in her mouth to her cubs (Kruuk 1976; Skinner and Ilani 1979). However, she does not regurgitate food for them. Striped hyaena cubs start eating meat when they are 30 days old, but continue to suckle (Rieger 1979a). Only one litter is raised at the same time in any one den (Mills 1978a).

4.4.3 Brown Hyaenas

Like striped hyaenas, female brown hyaenas are polyoestrous, with anoestrus during lactation. A female brown hyaena usually produces her first cubs when she is 2 years old, but the age of sexual maturity in males is unknown (Mills 1982a). Gestation lasts for 90 days (Eaton 1976a), and most mating takes place between nomadic, immigrant males and resident clan females (Mills 1982b; M.J. Owens and Owens 1984), the latter which defend the clan territory (D.D. Owens and Owens 1984). However, in the central Kalahari, the usual clan tenure of immigrant males is about 26 months. Consequently a specific male seldom sires more than two successive litters there. Immigrant males always dominate natal males in a clan (D.D. Owens and Owens 1984). Litter size varies from one to four, with a maximum of six young per litter (Eaton 1976a) and a mean of 2 (D.D. Owens and Owens 1984). A brown hyaena's cubs are born in a natal den which is usually an underground burrow, but sometimes a rocky or calcrete lair, an excavated aardvark burrow or even in clumps of bushes or tall grass (Eaton 1976a). The eyes are closed at birth, and open at eight days of age. Cubs have short hair, but of the same colour as the adults (Mills 1982a, 1990). The cubs are moved to the communal den of the clan when two to three months old, and are carried there by their mothers (D.D. Owens and Owens 1984).

Brown hyaena cubs in the southern Kalahari are suckled till they are three months old, and stay at the communal den till they are 15 months old (Mills 1982a). They may begin to spend time away from the den when 10 months old, but always return to it (Mills 1990). However, in the central Kalahari they may leave the den when only eight months old (Owens and Owens 1979a). Female brown hyaenas in the southern Kalahari will suckle cubs other than their own (allo-suckling), but they nevertheless show a clear preference for their own cubs (Mills 1990). Allo-suckling by brown hyaenas is also known from the central Kalahari (D.D. Owens and Owens 1984). Brown hyaena cubs are weaned when 10 to 12 months old (Mills 1982a, 1990; M.J. Owens and Owens 1984). In the central Kalahari, all the females in a clan provision the cubs, regardless of social rank and degree of relatedness. This favours cub survival in a semi-arid environment where a mother may have to travel long distances to feed herself and her cubs (Owens and Owens 1979a). In the Kalahari, the females provision young as distantly related as second cousins, but the resident clan males only help feed half-sibling cubs and do not provision cousins. These males also show little sexual interest in the females of their own, or of any other clan (Mills 1982b, 1989). Comparable discrimination in nonparental aid between relatives has to date only been known for primates (D.D. Owens and Owens 1984).

Brown hyaenas carry pieces of carcass to their own dens, offering it to the cubs at a much earlier age than in the case of spotted hyaenas (Mills 1989; Werdelin and Solounias 1991; Cooper 1993). However, the adults do not regurgitate food to the cubs (Owens and Owens 1979a). In the southern Kalahari, the mean distance over which food is carried to the cubs is 6.4 km, and the cubs start to eat meat as a food supplement when 12 weeks old (Mills 1989).

When the older clan members return to the den, brown hyaenas indulge in frequent and often prolonged greeting ceremonies. Adults and subadults visiting their dens in the central Kalahari scent-mark soon after arrival, and again before they leave (Owens and Owens 1979a). Usually only one female in a clan gives birth in a given year and there seems to be competition for breeding opportunities. Yet, even the most subordinate female gets a chance to breed and rear cubs unlike, for example, mongooses and social canids such as wolves and African wild dogs where only the dominant female breeds. The prolonged cycle of reproduction and development among brown hyaena cubs allows even the less dominant females to breed (D.D. Owens and Owens 1984).

Protection of the cubs at the communal den is of primary importance to brown hyaenas since the adults forage far and are often absent for long periods (Owens and Owens 1979a). Brown hyaena cubs can survive without food for several days (Hofer and East 1993c). They enter the interior parts of their dens through narrow entrance tunnels to escape from large predators such as wild dogs and lions (Owens and Owens 1979a). However, occasionally lions are known to kill brown hyaenas (Eloff 1973c; Mills 1990). On windy days when the cubs cannot hear well, they stay in the den and only venture out when the wind has subsided (Owens and Owens 1979a). Subadult brown hyaenas are between 15 and 30 months old. Physical development is complete at 30 months of age, when some individuals become reproductively active (Mills 1982d, 1990). Subadults help to rear the cubs by carrying food to the dens, but forage on their own (Mills 1990).

4.4.4 Aardwolves

A mated aardwolf pair in the Northern Province occupies a well-defended territory throughout the year. During the winter mating season, extra-pair copulation occurs in 40% of all pairs. The mating season starts when an adult male begins to make frequent scouting excursions into neighbouring territories, although initially it does not paste outside its own territory. Later, the male starts to paste in other territories too. When a female comes into oestrus, she pastes extra-limitally to such an extent that extra-limital scent-marking may become more frequent than pasting within her own territory. Extra-limital pasting does not displace the resident male, and therefore does not represent territorial extension. It is an advertisement of the qualities of the intruding male. Only the fitter males maintain their extra-limital pasting. Less fit males soon cease to paste in the territories of more aggressive resident neighbours (Richardson 1987c).

A week or so before the female's sexual cycle permits copulation, numerous males visit her, pasting frequently as they do so. A female paired with a less aggressive male will now also paste extensively outside the boundaries of her territory in an apparent attempt to get extra-pair males to visit her. However, at the time of oestrus, only the resident male, and perhaps an aggressive intruder, will remain with the female. Copulation lasts from 1 to 4.5 hours. Although the penis remains inserted all the time, there is no copulatory tie as in canids. The first ejaculation occurs after an hour or so, and then again at hourly intervals. During copulation, an aggressive intruder may displace a resident male. This means that cuckoldry occurs commonly. Copulation ceases when the female enters the den, forcing the male to dismount (Richardson 1987c). A litter of one to three cubs is born in summer after a 60-day gestation period. The eyes of the cubs are closed at birth (Grobler et al. 1984)

Cuckoldry is a form of sperm competition peculiar to animals that show extensive male parental care. As a result, a resident male may involuntarily rear another male's offspring as the latter male, the cuckold, may have inseminated the resident male's mate, even though the resident also copulated with her. It is generally believed that in monogamous species, as the aardwolf apparently is, the resident male should theoretically attempt to maximize its reproductive success by being promiscuous when the opportunity arises, while vigorously guarding its own mate when she is in oestrus. Promiscuity is widely practised by the aardwolf and leads to some males cuckolding their neighbours. With man and some birds, the aardwolf is one of only a few species of vertebrate in which cuckoldry is known to take place (Richardson 1987c).

Despite the obvious chances of cuckoldry, all males guard the cubs of their mates when they are confined to the den. This behaviour is unique among non-human mammals. It is not found in other monogamous carnivores such as foxes and jackals because of differences in the social structure of these animals. In some species of *Vulpes* and in jackals, the resident pair occupies a territory which they may share with their most recent young. When food is abundant, subadults remain in the parental territory but they do not breed. It is these extra defenders that probably prevent an intruder from gaining access to the resident adult female when she comes into oestrus. In the aardwolf, all subadults leave their natal territory, leaving only the resident male for an aggressive intruder to subdue. This is achieved mainly by prolific scent-marking, but the intruder may also fight with the resident male if necessary. The overtness of such extra-territorial behaviour suggests that cuckoldry is an established feature of aardwolf biology (Richardson 1987c).

An aggressive aardwolf male guards the cubs best. An apparently monogamous female may therefore be able to preclude the defection of her mate by mating with him as well as with an intruder (Richardson 1987c). However, if the female of a monogamous male is promiscuous, it has been predicted theoretically that he will defect and attempt to breed with an alternative female. This has been known to occur, but only rarely so because the resident male will probably still be raising his own cubs when the need to defect arises. Consequently, defection should only be possible on those rare occasions when a defecting male will have an alternative, receptive female available all to himself. Such a situation may arise through the natural death of the male of an adjacent pair during the short time frame when that male's mate is sexually receptive to mating. The potential for mate desertion is therefore low, but it always exists. Individual aardwolf males are likely to respond to local opportunities, whereas the behaviour of females is more likely to be determined by conditions in the population over the long-term (Richardson and Coetzee 1988).

Through cuckoldry, a female ensures that her cubs gain the higher quality genes of the intruder and that she produces young males which in turn become aggressive intruders elsewhere. Moreover, when fertilized both by her mate and by an intruder, she increases the genetic variability in future populations. As oestrus may last for three days at a time, multiple-mate matings are a real possibility in an aardwolf population (Richardson 1987c).

The nature of the diet of the aardwolf precludes the possibility that it can carry food back to its cubs, nor can it regurgitate food as the canids do. The only way for a male aardwolf to help raise the cubs is by defending the den, to whose vicinity the cubs are confined for their first three months of life. However, aardwolves frequently change den sites. While defending a den, a male aardwolf forages for only two to three hours just before dawn, but the female forages for most of the night. The cubs start to feed on termites when they are about three-and-a-half months old, but continue to suckle till they are five months old at most. Most of the cubs that die, do so as a result of food shortage in winter (Richardson 1987c).

4.5 Area Use, Movements, and Activity

In taxa such as the Carnivora, which show considerable interspecific and intraspecific variation, ecological factors alone may be of limited use in predicting spatial use and organization (Hofer and East 1993a). Most hyaenids occupy a large area, but the aardwolf is more limited in its space requirements. Even within a species, however, the pattern of area use may vary. Spotted hyaenas in the Serengeti, for example, move over vast distances, while those of the Ngorongoro Crater are more sedentary. The permanence of the prey base seems to play a major role in these differences (Kruuk 1970), as it does elsewhere in Africa. In the Serengeti National Park and the Savuti Marsh, a migratory prey base leads to large-scale movements, and even to the phenomenon whereby some hyaenas commute over the territories of several other clans to get to their food or water (Kruuk 1970; Cooper 1990b; Hofer and East 1993a,b,c; see below).

4.5.1 Spotted Hyaenas

The area used by a spotted hyaena clan generally overlaps little with that used by its neighbours. It is also defended against intruders and scent-marked. It is therefore a true territory (Mills 1982c). The territories of spotted hyaenas in the Serengeti are quite stable. They are defended, and their owners only commute to food when migratory prey leave their area. Hence, these hyaenas combine a residential life with frequent trips of several days away from their territories. During such forays, they may travel 40 to 80 km to forage in the area closest to the migratory herds (Hofer and East 1995). The intensity of commuting is related to the abundance of migratory prey in a given clan's territory (Hofer and East 1993c).

Spotted hyaenas, the most numerous of all carnivores in the Serengeti, defend permanent territories of approximately 55 km^2 per clan by scent-marking, vocal displays, boundary patrols and the aggressive expulsion of non-residents. Disputes with neighbouring clans commonly occur (Hofer and East 1993a, 1995). Each territory contains the communal den (Hofer and East 1993a), which is the focal point of a spotted hyaena's social system (Kruuk 1966; Hofer and East 1995).

The territories of spotted hyaenas in the Serengeti contain low densities (mean 3.3 animals per km^2) of resident herbivores throughout the year. However, during 26% of the year, the migratory herds are present and the density of prey then increases to 219 animals per km^2. By commuting to the migratory herds when necessary, the clan sizes of the spotted hyaenas in the Serengeti are maintained at levels higher than would be possible on a year-round basis. The feeding range of spotted hyaenas in the Serengeti is not part of territoriality (Hofer and East 1993a). This situation is thought to be unique to the Serengeti (Hofer and East 1993a), although commuting also occurs in the Savuti where a search for water is the primary incentive (Cooper 1990b).

The mean commuting distance of spotted hyaenas in the Serengeti is 40 km. Commuting destinations mirror the movements of the migratory game herds throughout the year. The commuters minimize contact with resident clans, and only go to areas where they are assured of finding migratory herds. The residents adjust their behaviour according to the context of an encounter with a commuter. They either ignore commuters in transit, or only respond aggressively at kills (Hofer and East 1993c). Female spotted hyaenas have commuting intervals of 3 to 4 days, while those of males are 6 to 10 days (Hofer and East 1993b). This may be because the females are often lactating and hence return earlier to attend to their cubs, although the latter can survive for several days without any food (Hofer and East 1993c).

In the Serengeti, spotted hyaenas spend 46 to 62% of the year commuting to food (Hoofer and East 1993b). The mean distance between simultaneously commuting members of the same clan is 13.68 km, which indicates that commuters from the same clan do not forage together. Individuals from the same clan do not commute to particular areas, and there is much overlap between the commuting destinations of different clans. Subadult, but still dependent, cubs may accompany their mothers on commuting trips. In this way they probably learn the major routes and become familiar with reliable destinations. Apart from these commuting trips, the spotted hyaenas of the Serengeti also make other foraging trips, but these are all within the boundaries of the territory (Hofer and East 1993c).

It has been suggested that the development of a food foraging commuting system among Serengeti spotted hyaenas is based upon two sets of factors. The first set involves five main areas of adaptive specialization, which occur in all spotted hyaena populations and permit a large foraging radius. The most important of these specializations is the use of dens accessible only to the cubs. This reduces the risk of them being preyed upon when the mother is absent, and provides a focal foraging area. Furthermore, the use of communal dens reduces the probability that the cubs will be unguarded while their mothers forage. In addition, and as already mentioned, spotted hyaena cubs can survive for several days without any nutrition. Morphological and physiological characteristics endow spotted hyaenas with high mobility, and hence a large foraging radius. Finally, the potential for overheating during extended periods of exercise could possibly be countered in the spotted hyaena by cooling of the arterial blood supply to the brain by countercurrent heat exchange with venous blood cooled by evaporation. Although such a system is known to occur in some mammals, it has not yet been described for hyaenids. However, the large rostrum of the hyaenids may well make such a system possible as in some canids. The second set of factors involves aspects which are unique to the Serengeti ecosystem. One of the most important of these is the extended prey-poor periods which constrain the lactating females, especially as they have such an extended lactating period. Furthermore, commuters are tolerated behaviourally by the other spotted hyaenas through whose territories they pass. The benefit of this flexibility in territorial behaviour is that, while the commuters are able to forage widely in search of food, the residents of the area can take over the kills at relatively little risk to themselves. This complex array of behavioural, morphological and physiological specializations enables the spotted hyaenas of the Serengeti to forage effectively (Hofer and East 1993c).

Relative to those of the more mesic Ngorongoro Crater, spotted hyaena clans in the Masai Mara National Reserve occupy territories which are widely separated by buffer zones in which prey is scarce. Hence, the maintainance of territorial boundaries is rare (Frank 1986a). In the even more arid and prey-poor central Namib Desert, maintenance of territories is absent. This is because clan sizes are small, home ranges are large and the temporal distribution patterns of spotted hyaenas prevents overlaps (Tilson and Henschel 1986). It is therefore clear that territorial behaviour in spotted hyaenas is flexible and in part dependent on local ecological conditions. In this context, aridity, and its concurrent prey scarcity, is a major ecological factor.

The local abundance of ungulates in the Masai Mara area largely determines the distribution pattern of its spotted hyaenas. Because prey-rich, shortgrass areas are relatively small and scattered throughout a large area which generally supports little game, the boundaries of the clan ranges there tend to be diffuse. Clan ranges in the Masai Mara area form a mosaic of coreareas separated by areas of no-man's land. In contrast to the well-defined territorial boundaries and complex behavioural patterns in the Ngorongoro Crater, the spotted hyaenas of the more arid Masai Mara area rarely show such behaviour (Frank 1986a).

The Kruger National Park and its adjacent private nature reserves harbour a substantial prey base. Yet, the territory size of a spotted hyaena clan is considerably larger there than in the Serengeti (Henschel and Skinner 1987, 1991). In the Timbavati Private Nature Reserve, a spotted hyaena territory is some 25 km^2 (Mills 1978a), while in the Kruger National Park it is 130 km^2 (Henschel and Skinner 1987). Territorial activities by resident spotted hyaenas in the Kruger National Park are directly proportional to the intrusive pressure of neighbours. Resident males on their own fail to evict intruding females, and even sometimes follow such females to their own clans as prospective mates. When clan size was reduced by 25% on one occasion, the ability of the clan to defend disputed parts of its range declined too. As a result, larger neighbouring clans appropriated part of the range (Henschel and Skinner 1991).

Compared with the prey-rich Serengeti ecosystem and the relatively prey-rich Kruger National Park, the southern Kalahari has a low prey base for spotted hyaenas. Consequently, the clans of the Kalahari are small (three to 13 adults and subadults) and occupy large territories (mean: 1095 km^2) (Mills and Gorman 1987). The smallest spotted hyaena territory in the southern Kalahari is 55 times larger than the largest one in the more mesic Hluhluwe Game Reserve (Mills 1990). In the Kalahari Gemsbok National Park, the dry Nossob riverbed is central to most of the territories of the spotted hyaena. This riverbed is prey-rich when compared with the adjacent dune areas, and most spotted hyaena movements occur along it (Mills and Gorman 1987). The degree of territorial overlap is also greatest where food is most abundant (Mills 1978a). The actual territory size of a specific clan is apparently determined by the manner in which its food resources are spaced. Interactions between different spotted hyaena clans are rare in the Kalahari (Mills 1990).

Migrating prey and the sparse distribution of water in the Savuti cause the spotted hyaenas there to commute to neighbouring territories to seek water. However, they return to their own territories to hunt. Disputes between clans usually occur in areas of high prey density and often involve food. When fights do occur, they are noisy but injuries are rare. In areas where food and water resources are evenly spaced, the territory sizes of the spotted hyaena clans are similar, even in different habitats. In the Savuti, each clan has a permanent water source in its territory (Cooper 1990b).

The mean range size of 360 km^2 of spotted hyaenas in the Etosha National Park is also large when compared with those of East Africa (Gasaway et al. 1989). However, it is still much smaller than the ranges in the Kalahari (Mills 1990). The spotted hyaena clans in Etosha live in discrete areas which provide the year-round requirements of each clan. The large size of the ranges there is attributed to a patchy and variable food and water supply. However, no territorial defence is found in those clans. Hence these areas of use are more probably ranges than territories in the strict sense of the definition (Gasaway et al. 1989).

Elsewhere in Namibia, the spotted hyaenas of the central Namib Desert use foraging areas which are large (mean: 570 km^2; range 383 to 816 km^2) (Tilson and Henschel 1986), and exclusive to each clan (Tilson et al. 1980). Intergroup encounters as found in East Africa are absent. The small clan sizes and large range-use of the Namib spotted hyaenas appear to be an adjustment to a dispersed and depauperate prey-base, and to a limited water resource which lies along the Kuiseb River Canyon. Each spotted hyaena clan's range in the central Namib is centred on specific sites found along this canyon. (Fig. 4.2; Tilson et al. 1980; Tilson and Henschel 1986). These sites contain breeding dens, daytime dens, daytime lairs, and permanent water sources (holes, seeps or springs). The core-areas of adjacent ranges are 20 to 35 km apart. The mean range size is 20 times larger than that of spotted hyaena clans in the Ngorongoro Crater. The boundaries between adjacent ranges are separated by a resource-poor no-man's land of as much as 15 km across. The maximum linear distance from the core area to the outermost food resource utilized by one clan is 30 km. Carcasses further away than this are not utilized, probably because a round-trip of 60 km per night is the maximum possible there (Tilson and Henschel 1986). In contrast, spotted hyaenas in the southern Kalahari can travel up to 40 km from their dens to a food resource, and return the same night (Eloff 1964). Probably because of their range separation, there is no interaction between different clans in the Namib Desert, and range boundaries are marked infrequently. When a water source dries up, a clan either disperses or relocates to a new core-area in its range, provided that another water source is available. Thus, as in other carnivores, the spatial organization of spotted hyaenas in the central Namib Desert varies with the availability and spatial distribution of water and food as key resources (Tilson and Henschel 1986). These are both affected by the aridity of the region.

Dens are the centre of social activity in a spotted and brown hyaena clan. All members of a clan visit the den where they socialize with the cubs and other clan members (Mills 1990). Where warthogs and spotted hyaenas occur together, they often share burrows. This is possible because their times of activity are different (Deane 1962). The occupancy of a den in the southern Kalahari is much less than the six months, or even years, which the spotted hyaenas of the Kruger National Park spend there. Flea infestations may explain the brief occupancy of dens in the Kalahari. This is particularly the case in large dens, while new dens are free from fleas. At least 50% of the dens of spotted hyaenas in the southern Kalahari are taken over from porcupines and aardvarks. Most dens are used for one period only, but some may be occupied for years. Spotted hyaenas in the southern Kalahari are predominantly nocturnal. This helps them to avoid competition from vultures and also conserves water (Mills 1990). Spotted hyaenas are less active than brown hyaenas in the southern Kalahari, because the latter need to move around

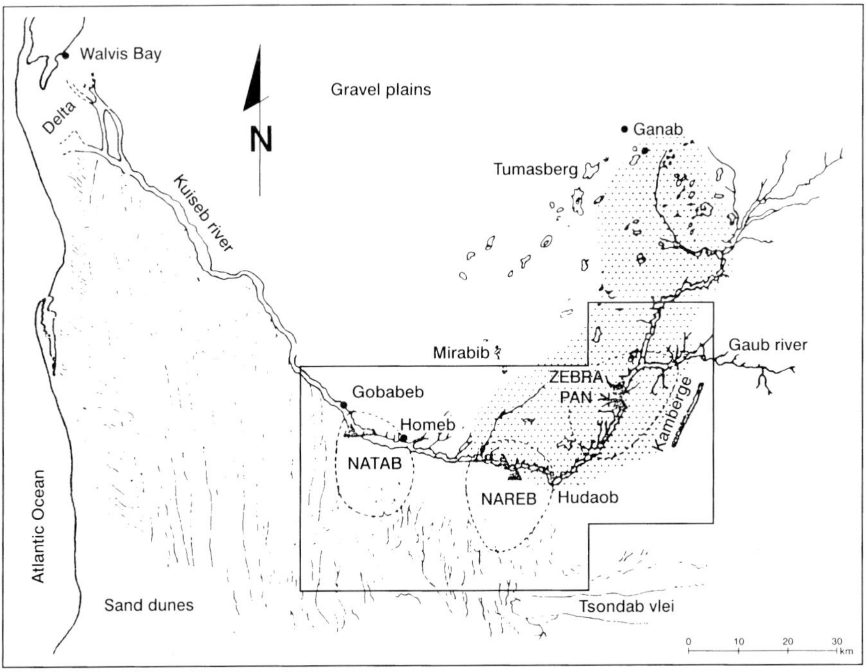

Fig. 4.2. The spacing of three spotted hyaena clans at Natab, Nareb and Zebra Pan along the Kuiseb River Canyon in the central Namib Desert of Namibia. Foraging ranges are outlined in dashed lines; shaded area (stippled) shows the general distribution of Hartmann's mountain zebra (Tilson et al. 1980)

constantly in search of food. The maximum distance travelled per night by a spotted hyaena in the southern Kalahari is 69.1 km in a little over 10 h, at a mean speed of 6.7 km/h. The mean daily distance travelled is 27.1 km. During the day a spotted hyaena will rest in an aardvark or porcupine burrow (there are usually no warthogs in this region), under a shady tree (usually a densely foliated shepherd's tree) or even in a waterhole (Mills 1990). However, according to an earlier warden of the Kalahari Gemsbok National Park, one clan may have moved at least 80 km in a single night while hunting (Eloff 1964).

Male spotted hyaenas in the Masai Mara National Reserve disperse from the natal group at the communal den around puberty, when they are about 2 to 2.5 years old (Frank 1986a). The movements of spotted and brown hyaenas in the southern Kalahari are dictated by the distribution of their food and the location of the breeding dens. The communal den is the focal point of spotted and brown hyaena society and spotted hyaenas will return to their den irrespective of their foraging success (Mills 1990). The communal den of spotted hyaenas in Savuti is not used to feed the cubs, but rather to enable the cubs to socialize with the clan (Cooper 1993).

Spotted hyaenas reach their highest densities in areas such as the Serengeti ecosystem with its flat, open grasslands with a few scattered, or no trees, and a large prey resource (Mills 1978a). Probably the highest density - one individual per 0.6 km^2 (170 per 100 km^2) - occurs in the Ngorongoro Crater. In the Masai Mara National Reserve the highest density is one individual per 8.3 km^2 (12 per 100 km^2) (Frank 1986a). The density of any carnivore population is difficult to estimate accurately. Nevertheless, it is estimated that there are twice as many brown as spotted hyaenas in the Kalahari Gemsbok National Park. The densities there are estimated at one individual per 100 to 125 km^2 (0.8 to 1 per 100 km^2) for spotted hyaenas, and one individual per 55 km^2 (1.8 per 100 km^2) for brown hyaenas (Mills 1990). This is considerably lower than the density of one individual (adults and subadults) per 20 km^2 (5 per 100 km^2) for spotted hyaenas in the Etosha National Park, where clan sizes are larger, and there is more food and water than in the southern Kalahari (Gasaway et al. 1989).

4.5.2 Striped Hyaenas

The social behaviour of the striped hyaena appears to be similar to that of the brown hyaena. In the Serengeti ecosystem a striped hyaena has a small home range of some 44 km^2 for a subadult female and 72 km^2 for an adult male (Kruuk 1976; Mills 1978a). The range of the striped hyaena does not appear to be a defended territory (Kruuk 1976).

Striped hyaenas spend most, and especially the first part of the night, foraging or moving between foraging sites. Daytime lairs under rock overhangs are often used to rest in (Kruuk 1976). In the Negev Desert, some striped hyaenas regularly follow footpaths (Macdonald 1978), but others do so only infrequently, as in East Africa (Kruuk 1976). Regular paths are useful only if they lead to a localized food resource as is the case in Israel. For the most part, a striped hyaena will arrive at a feeding site, eat, sleep, and then depart (Macdonald 1978). In India, striped hyaenas rest by day in rocky ravines, or in burrows which they dig themselves (Kruuk 1976).

Striped hyaenas prefer dry, open and often rocky habitat with thick bush, where there is a low incidence of large ungulates as in the Negev Desert (Mills 1978a). The density of striped hyaenas in that desert is one individual per 300 km^2 (0.3 per 100 km^2) compared with one individual per 58 km^2 (1.7 per 100 km^2) in the Serengeti (Bouskila 1984).

4.5.3 Brown Hyaenas

Brown hyaenas commonly go underground by day to rest in dens taken over from other animals, or which they have dug for themselves (Skinner 1976; Mills 1990). They move around considerably at night, especially when feeding on one small food item after another (Mills 1990). The adults do not sleep at the den during the day (Owens and Owens 1979a). In the southern Kalahari, brown hyaenas, unlike spotted hyeanas, often have one or two minor dens 5 to 500 m away from the main den. As with the spotted hyaena, the period of den occupancy by a brown hyaena in the southern Kalahari is short, again probably because of flea infestations. Brown hyaenas tend to den in the same vicinity, but not necessarily the same den, for years. The short duration of occupancy of a given den in the southern Kalahari differs from the long period of den use found in the brown hyaena in the central Kalahari (Mills 1990). The reasons for this difference are not clear. Brown hyaenas travel a mean distance of 31.1 km per night with a maximum of 54.4 km in almost 14 hours at a mean speed of 3.9 km/h (Mills 1990). Adult brown hyaenas in the southern Kalahari are active for 43% of the 24-h cycle (Mills 1990). In the central Kalahari, two activity peaks occur per night - early in the evening soon after dusk and late at night just before dawn (Owens and Owens 1978).

The brown hyaenas of the Kalahari Gemsbok National Park, which live in the interior dunes and the open, sandy savanna away from the riverbeds, show a random pattern of movements with several localized peaks. The food there is distributed randomly too. Those living in the dry Nossob riverbed, an area of high ungulate density which provides an abundance of food for brown hyaenas through the kills of other large predators, mainly follow the riverbed in their movements (Mills and Mills 1982).

Of all the large carnivores in the southern Kalahari, the brown hyaena is the most common. This is because its habitat preference is for dry, open country (Mills 1978a). It is twice as abundant there as the spotted hyaena (Mills 1990). Nevertheless, the brown hyaena tends to avoid areas of high spotted hyaena density (Mills and Mills 1982).

Brown hyaenas that live in the southern Kalahari have large, overlapping ranges of 235 to 540 km² (mean 308 km²). The sizes of these ranges depend upon the quality of the food resources available (Mills 1978a, 1983; Mills and Mills 1982). There is never more than 20% overlap between adjacent ranges (Mills 1983). In the central Kalahari, a brown hyaena actively defends its range. Consequently, it is considered a terrritory (Owens and Owens 1979a), although these ranges also overlap extensively as in the southern Kalahari (Owens and Owens 1978). Brown hyaenas in the Namib Desert live in much larger territories than in the Kalahari. The size of a territory there is dictated by the available food, which is dictated in turn by the extreme aridity of the region (Mills 1990).

4.5.4 Aardwolves

The aardwolves living in the Serengeti ecosystem do so in pairs in territories of some 1.5 km². Each pair has a den at the approximate centre of its range, and the boundaries are scent-marked (Kruuk 1972b). In the Northern Cape Province, a mated pair of aardwolves lives in a territory which is defended against intruders (Richardson 1987c, 1990). In one bushveld area in South Africa, the mean area used by an aardwolf is 1 km² (range: 0.8 to 1.3 km²) (Skinner and Van Aarde 1986). However, when mating starts, the adult male begins to make frequent scouting excursions into neighbouring territories (Richardson 1987c). Feeding requirements influence territory size, but breeding requirements are even more important in this regard. A mated pair shares the same territory exactly (Richardson 1987c).

Aardwolves are mainly nocturnal since their main food resource, the long-nosed harvester termite *Trinervitermes trinervoides*, only becomes active at night. As indicated later, however, the aardwolf becomes more diurnal in winter to supplement its diet (Fig. 4.3). In winter, the aardwolves of the central Namib Desert emerge from their burrows 1.5 h before sunset. In summer they emerge at or after sunset, returning to the burrow before dawn in both seasons. Several different burrows are visited at night for periods of rest while foraging (Nel and Bothma 1983). In winter, 88% of all activity takes place between emergence from the den and midnight, but in summer the aardwolves of the Namib are active throughout the short nights (Bothma et al. 1984).

The foraging distance of a female aardwolf in one arid bushveld area of South Africa varied from 1.5 to 9.1 km per night (mean: 4.2 km). On the western grasslands, a relocated aardwolf once moved 9.7 km within 3 days back in the direction of its orginal capture site, while another relocated individual was found 3.2 km from its original capture site, 5 months later (Bothma 1971b).

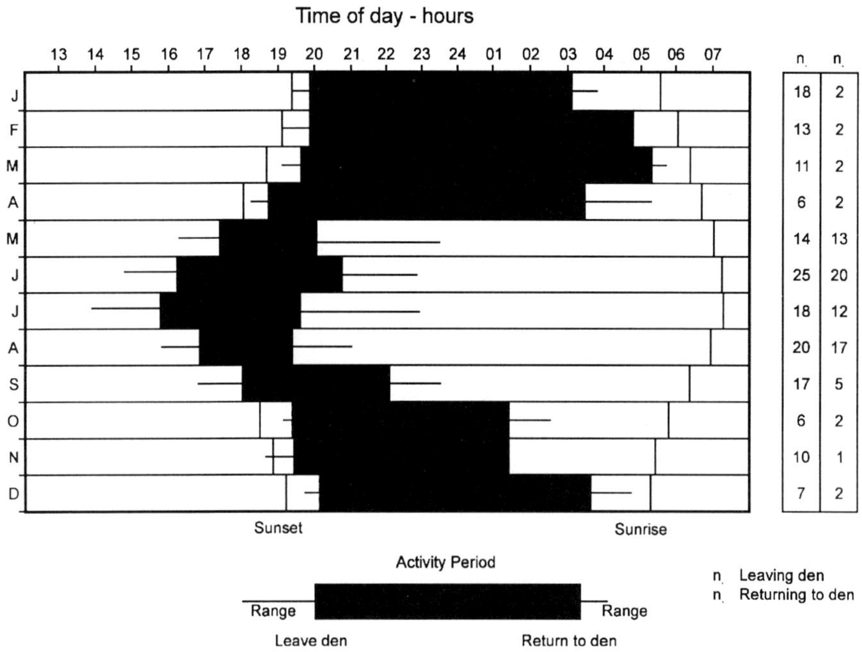

Fig. 4.3. Aardwolf activity periods in the Northern Cape Province of South Africa during the year and relative to sunset and sunrise (Richardson 1987b)

The aardwolf prefers open areas and avoids densely wooded habitats such as those found in parts of the Serengeti Plains (Mills 1978a). In the arid bushveld areas of central South Africa there is a density of one aardwolf of all ages per 0.4 km^2 (250 per 100 km^2) and one adult per 1 km^2 (100 per 100 km^2) (Skinner and Van Aarde 1986).

4.6 Feeding Ecology

The drive to acquire food is a vital force in the life history of all animals. Individuals must obtain sufficient nutrients to remain alive, to reproduce successfully and to raise their young (Mills 1990). The hyaenids vary from pack hunters to solitary foragers (Houston 1979). They are unique in the development of bone-crushing teeth, except in the case of the aardwolf which has severely reduced dentition (Mills 1990). This is a wide array of foraging methods to find in a family with only four living species. Most carnivores will scavenge food whenever it is available, but hyaenids do so more often than others. Consequently, they show several adaptations for doing so (Houston 1979).

Amongst the subfamily Hyaeninae, the dentition of the spotted hyaena is best adapted for meat eating, followed by that of the brown hyaena and then the striped hyaena (Mills 1978a, 1990).

4.6.1 Food

Spotted Hyaenas. A spotted hyaena will eat almost anything that is remotely digestible (Bearder 1977). Until fairly recently, the view was commonly held that the spotted hyaena mainly scavenges for food (Houston 1979). Yet, it has long been known that spotted hyaenas do not only scavenge (Johnston 1884) although they are major scavengers (Bertram 1979). The true extent of such hunting was first reported by Eloff (1964). Scavenging requires an animal to travel long distances with low energy expenditure (Bertram 1979). This characteristic fits spotted hyaenas well. They digest large amounts of bone, and their canines are specialized to crush large bones (Anyonge 1996).

Differences in the availability of prey lead to differences in the diet of spotted hyaenas in different regions. This is seen, for example, in the spotted hyaenas of the Serengeti National Park compared with those of the Ngorongoro Crater (Kruuk 1966, 1972b). In the Serengeti, only 58% of the carcasses on which spotted hyaenas feed are killed by the hyaenas. In comparison, 96% of the carcasses eaten by spotted hyaenas in the Ngorongoro Crater are killed by them (Mills 1978a). Spotted hyaenas feed mainly on large ungulates. Of all predator kills in the Serengeti National Park and the Ngorongoro Crater combined, 82% are killed by the spotted hyaenas, 11% by other predators and in 7% the cause of death is unknown (Kruuk 1966). Even in the use of bone there are regional differences in spotted hyaenas. Those living in East Africa carry bone to their dens more often than do those in South Africa (Sutcliffe 1970).

The dietary spectrum of spotted hyaenas in the Kruger National Park is much wider than that of lions or any of the other large predators (Mills and Biggs 1993). It varies with habitat, season and ungulate abundance (Smuts 1979; Henschel and Skinner 1990). Impala (67.9% of all food), Burchell's zebra and blue wildebeest form the main prey of spotted hyaenas there. In the lambing season, 87.5% of all food eaten by spotted hyaenas in the Kruger National Park consists of impala, of which 92% are lambs. This temporary dietary focus during the lambing season is similar to what is found in the Ngorongoro Crater when the wildebeest calve. In the Kruger National Park, most of the impalas eaten are also killed by spotted hyaenas. Consequently, generalizations about the diet of the spotted hyaena should be avoided (Smuts 1979).

Carcasses are scavenged to some extent by spotted hyaenas in the Kruger National Park, but more than 50% of their food is hunted (Henschel and Skinner 1990; Mills and Biggs 1993). Even the occasional case of cannibalism is known (Pienaar 1969). Scavenging is predominant during periods of drought when the impala, kudu and warthog are particularly vulnerable to starvation (Henschel and Skinner 1990). When scavenging from other predators, lion kills form an important source of food (Bearder 1977). However, the hyaenas usually only approach a carcass after the lions have finished feeding (Mills and Biggs 1993). By regurgitation, spotted hyaenas can also deal with relatively indigestive material, such as grass which may be consumed while feeding (Bearder 1977). However, probably most of the plant remains found in spotted hyaena scats are from the rumen contents of their ungulate prey (Henschel and Skinner 1990). The diet also includes dead lions, and even the occasional civet and wild cat (Smuts 1979). As the most social of all living hyaenids, spotted hyaenas feed mainly on large carcasses which will provide enough food for several hyaenas at one time (Wilson 1975).

Where large ungulates are relatively scarce, as in the southern Kalahari, the spotted hyaena expands its diet to include any kind of mammal. It also eats birds, reptiles, amphibians, invertebrates, and readily eats plant material. The latter includes wild fruits such as the tsama melon and wild cucumber *Cucumis africanus* (Mills 1978b, 1990). Of the mammals, large ones of 500 kg or more and small ones of 12 to 80 kg are more often eaten than medium-sized ones. Gemsbok calves (43.3% of all kills) are an important food resource. Of gemsbok killed in the southern Kalahari, 80.8% are calves in all stages of growth (Mills 1990). Although adult gemsbok are often hunted, they can fend off and even kill spotted hyaenas (Eloff 1973a,c; Mills 1990). The blue wildebeest is the prey killed second most often in the southern Kalahari, but calves there form a smaller proportion (66.7%) of the wildebeest killed, and the age of the calf is irrelevant. This is in contrast to what occurs in the Serengeti and the Ngorongoro Crater (Mills 1990).

Springbok lambs are available for a short time only in the southern Kalahari. Although smaller mammals, such as the steenbok, porcupine, hares and the springhare are common, they are rarely preyed upon by spotted hyaenas. Occasionally, a spotted hyaena will eat a few insects, but not as often as a brown hyaena does. Rodents are also occasionally used as food, as are ostrich chicks. The aardvark is not eaten. It easily escapes predation by darting into a burrow. Smaller carnivores such as polecats, bat-eared foxes, black-backed jackals and honey-badgers are chased but not killed. This may be more play than hunting (Mills 1990).

Fat-rich termites are opportunistically licked up by spotted hyaenas in the Savuti. There, however, half the meals found by the spotted hyaenas are carrion, mostly bones. However, these bones are often ignored as food, or when eaten, only incompletely so. In contrast to the southern Kalahari, the spotted hyaenas of the Savuti occasionally kill and eat springhares, and even some of the smaller carnivores, although the flesh of such carnivores is often rejected. In Savuti also, neither the abundant blue wildebeest nor the Cape buffalo are important prey for spotted hyaenas, possibly because of the defensive action of the large blue wildebeest bulls, and the sheer size of a buffalo. The successful defence of blue wildebeest bulls against predator attack in Savuti is a local phenomenon. Elsewhere, as in the southern Kalahari and East Africa, blue wildebeest are often killed by spotted hyaenas (Cooper 1990a).

In the Etosha National Park, the spotted hyaena is primarily a hunter and kills 75% of its own food, with springbok its primary target. Unlike elsewhere, Burchell's zebra is not hunted intentionally, but an occasional foal may be killed by spotted hyaenas in central Etosha if it is less than 3 months old. The high reliance on hunted food in Etosha is the result of a low incidence of carrion because the other large predators consume most or all of their kills. The reason why they do so compared with other regions is unclear. The main source of carrion is provided by sporadic outbreaks of anthrax when many ungulates die. Furthermore, the arid and hot climate leads to the rapid desiccation of carcasses which then become of no interest to the spotted hyaenas (Gasaway et al. 1991).

Adult gemsbok (79.6% of 595 scats) are killed regularly by spotted hyaenas in the central Namib Desert where, in most areas, the spotted hyaena is the largest carnivore (Tilson and Henschel 1986; Mills 1990). However, some clans kill Hartmann's mountain zebras *Equus zebra hartmannae* more often than gemsbok. This is because of local differences in abundance. In this region, the diet selection of spotted hyaenas depends upon their foraging range and the availability of prey. Old gemsbok and zebras are most often killed (Tilson et al. 1980). Except for a few leopards (Skinner et al. 1992), other large carnivores are absent in this area. Consequently, carcasses are rare. Smaller antelope such as springbok, steenbok and klipspringer *Oreotragus oreotragus* only form 2.2% of the diet of spotted hyaenas in the Namib, and a case of cannibalism is known from there. Hartmann's mountain zebra only occurs in the upstream areas of the Kuiseb River Canyon. This canyon is the focus of the foraging areas of all the Namib spotted hyaenas (Fig. 4.2.). Hence these zebras only occur in the diet of one clan. Furthermore, the Kuiseb River Canyon and its adjacent areas are poor in mammal abundance and diversity, placing a greater reliance on gemsbok as prey for spotted hyaenas. Aggressive male gemsbok often injure each other. This reduces their ability to escape when hunted by spotted hyaenas. In times of drought, the death of ungulates due to starvation provides a major source of food for spotted hyaenas. In contrast, hunting provides most of their food in times of abundant rain (Tilson et al. 1980).

Striped Hyaenas. Striped hyaenas are effective scavengers, but relatively poor hunters (Mills 1990). The food of a striped hyaena is mainly carrion, especially the bony remains of carcasses (Rieger 1979a). In the Serengeti National Park and the Arabian Peninsula, the striped hyaena is more omnivorous than the spotted hyaena. It eats insects, wild fruits and small vertebrate prey in addition to carrion (Kruuk 1976; Nader 1996). In Serengeti, the fruit most often eaten is that of the wild plum *Balanites aegyptiaca*, while those of the marula *Sclerocarya birrea* are never eaten, although they are abundant and nutritious (n = 42 scats) (Kruuk 1976; Palgrave 1993). Especially impalas and

Thomson's gazelles, but also Cape hares, bat-eared foxes, cheetah cubs, reedbuck and dik-diks *Madoqua kirkii* are hunted or scavenged (Kruuk 1976). In many other areas, striped hyaenas scavenge human waste (Kruuk 1976). In the Negev Desert, the striped hyaena is primarily a scavenger, but also eats plant material such as desert dates (Bouskila 1984).

Brown Hyaenas. The brown hyaena is well-adapted physiologically, morphologically and behaviourally to exist in an arid environment (Mills 1977). A scavenger, it concentrates around the kills of other predators. It is an opportunistic carnivore with a generalized diet (Eaton 1976a) and eats ostrich eggs when available (Skinner 1976; Mills 1990). In the southern Kalahari, 96% of its vertebrate food is scavenged (Mills 1990). It occasionally supplements its diet with insects, wild fruits, bird's eggs and an occasional small vertebrate which it hunts (Mills 1978b). Of its faeces, 60.1% contain insect remains, and 19.6% reptile remains (Mills 1990). The wild fruits eaten are primarily the tsama melon and the gemsbok cucumber *Acanthosicyos naudianus*. The tsama melon is utilized for a longer period than the gemsbok cucumber, which is more susceptible to frost. These fruits are also eaten when the brown hyaena drinks water, and are therefore nutritional as well as a source of moisture (Mills 1978b). Wild fruits comprise 29% of all food items eaten by brown hyaenas, but are of low nutritional value (30 to 100 kJ per 100 g) (Mills 1982c, 1990). Tsamas and gemsbok cucumbers are, however, rich in vitamin C (Mills 1990). Other wild fruits eaten include those of the raisin bush *Grewia flava* and the wild cucumber. The nabba *Terfezia pfelii*, a wild fungus which is also known as the Kalahari truffle, is also readily eaten when available. Brown hyaenas in the Kalahari may be important agents for the distribution of fruit seeds in their dung. Brown hyaena cubs, however, eat fewer fruits than adults (Mills 1978b). Of the mammal remains scavenged, blue wildebeest, gemsbok, springbok, steenbok and small canids are eaten most (Mills 1982c).

Ostrich eggs are prized food items. They are eaten regularly by brown hyaenas in the Kalahari , but rarely so by lions or spotted hyaenas, although these should have no difficulty in biting them open as brown hyaenas do. Single ostrich eggs are eaten routinely by brown hyaenas, but brown and spotted hyaenas will also raid an ostrich nest. The brown hyaena will successfully eat the eggs, while the spotted hyaena attacks the brooding birds and only attempts to eat the eggs. A brown hyaena easily bites open the top of an ostrich egg to lap up the contents, and even what spills onto the sand. In contrast, a spotted hyaena has difficuly in biting open an ostrich egg, but often kicks an egg against another one in a nest to crack it ,whereupon it can be opened easily (Mills 1990).

Many of the items eaten by brown hyaenas in the Kalahari are so small that they only provide enough food for one animal at a time. The diet of brown hyaenas in the southern Kalahari is generally similar to that in the central Kalahari and elsewhere in South Africa. However, it differs vastly from the more restricted diet of brown hyaenas in the Namib Desert (Mills 1990). In the dry season, large mammal remains are only eaten in 17% of cases by brown hyaenas in the central Kalahari, but they are eaten extensively during the wet season. Of vertebrate food eaten, only 16% is killed by the hyaenas themselves. This includes small mammals, birds and reptiles(n = 258 scats) (Owens and Owens 1978).

The diet of the brown hyaena in the Namib Desert is strongly influenced by the presence of seal rookeries along the coast. These rookeries provide an abundant food source. The brown hyaenas of the Namib live mainly along the coast, and 75% of their scats (n = 68) contain seal remains. These brown hyaenas primarily forage for dead seal pups. In contrast, the spotted hyaenas of the Namib mostly forage inland and their diet contains an abundance (94.5% of 36 scats in one study) of gemsbok remains (Skinner and Van Aarde 1981; Stuart and Shaughnessy 1984). In addition to seals, the brown hyaenas along the Namib Desert coast also eat seabirds (n = 120 scats). These are jackass penguins *Spheniscus demersus* (Siegfried 1984), which are probably scavenged. In the Judean and Negev Deserts, the diet of the brown hyaena is similar to that of the striped hyaena (Skinner and Ilani 1979).

Aardwolves. It is now well known that the aardwolf is a specialized feeder which exploits various ecologically similar species of the harvester termite genus *Trinervitermes* extensively. The extent of the impact on colonies of termites is evident in that the stomach of a single aardwolf may contain as many as 124 110 individual termites at a given time (n = 7 stomachs) (Bothma 1971d). A single aardwolf can lick up as many as 300 000 termites in a single night (Richardson and Levitan 1994). Occasionally, however, aardwolves will eat other types of food, including other species of termite, dung beetles, flying ants and even an occasional leopard tortoise which may well have been scavenged (Bothma 1971d). In the Serengeti ecosystem, there is a high degree of selectivity for *Trinervitermes bettonianus*. This termite is present in 97.5% of all scats and, in 46% of cases, it is the only food present. During the rains, army ants *Dorylus* spp. and other invertebrates such as *Odontotermes badius* are also eaten. Whenever termites are eaten, both workers and soldiers are taken, but the soldiers are eaten less often and sometimes not at all (Kruuk 1972c). It is possible that the aardwolf's absence from many parts of Africa may be due mainly to the relative scarcity of termites which forage in the same way as East Africa's *Trinervitermes bettonianus* (Fig. 4.4; Kruuk 1972c; Richardson 1987a,b).

That environmental conditions should influence the diet of an aardwolf, regardless of its common dietary specialization, is clear from the Northern Cape Province. The winter nights are often so cold (< 18 °C at sunset) that they prevent the main food of the aardwolf *Trinervitermes trinervoides* from becoming active. Although aardwolves there feed mainly on this species for most of the year, they switch their diet in winter to another harvester termite

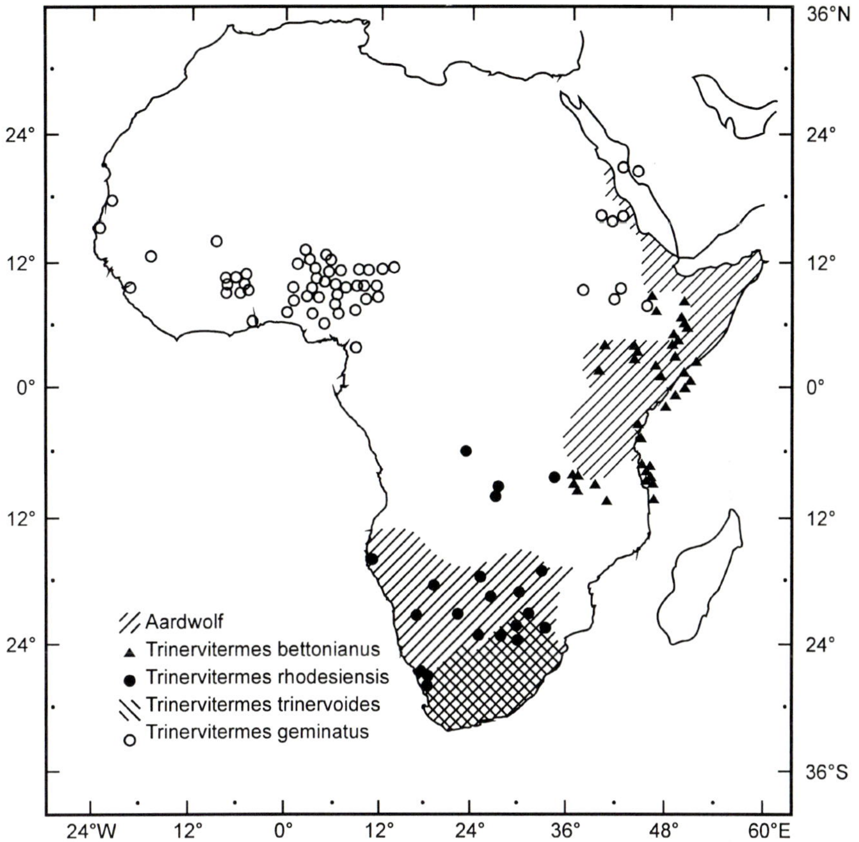

Fig. 4.4. The present distribution of the aardwolf in Africa, and the distribution of four species of surface-foraging Trinervitermes termites (Richardson1987a)

Hodotermes mossambicus which is mainly active by day during the winter. With the reduced food resources available to them in the winter, aardwolves may lose up to 20% of their body mass then (Richardson 1987b,c).

There seems to be no significant correlation between rainfall and termite abudance in the central Namib Desert. However, there is a switch in the diet of aardwolves to relatively more *Hodotermes mossambicus* in the cold winter, in the same way as in the Northern Cape Province (see above). In the Namib Desert, 96% of the food of an aardwolf consists of *Trinervitermes trinervoides* (n = 79 scats). Some other insects are eaten in small quantities too (Bothma et al. 1984).

Among the myrmecophagous mammals, the aardwolf is unique in that it feeds almost exclusively on ecologically similar species of a single genus,

Trinervitermes. Yet it posesses none of the obvious external features such as a long snout, and powerful limbs and claws which are characteristic of this feeding style. The species of the genus *Trinervitermes* eaten are all nocturnal, surface-foraging insects, active for most of the year and with a highly effective chemical defence. This is secreted from the frontal glands of the soldiers. It is a noxious, terpene-based substance which is sprayed onto an attacker. The aardwolf's most significant dietary adaptation seems to be that it can tolerate this secretion. Consequently it can feed effectively without having to dig for its food (Anderson et al. 1992; Richardson and Levitan 1994).

As an aardwolf licks up termites from the ground, it sometimes ingests up to 50% debris unintentionally (Bothma and Nel 1980; Richardson 1987a). Its metabolic ratio is only 70% of what it should be for an average mammal of the same body weight (Richardson 1987a). The aardwolf, however, probably did not evolve into its specialized diet slowly, because its low reproductive rate would not have allowed successful competition with other myrmecophores of the time. Most myrmecophores are specialized for eating any small arthropod, and not for specific species. Hence the ancient aardwolves must have been physiologically pre-adapted to tolerate the noxious secretions of their principal diet (Richardson 1987a; Richardson and Levitan 1994). The subfamily Protelinae is recent among carnivores, having first evolved only 1.5 million years ago from civet-like viverrid ancestors. The civet is notorious for its ability to eat food items that are toxic, or at least distasteful to most other mammals. From this departure point, dietary specialization could have followed naturally as other myrmecophore adaptations, such as the broad tongue and palate, large salivary glands and well-developed senses of smell and hearing, emerged. The penalty incurred from this specialization is that the aardwolf is absent from areas where there are no surface-foraging *Trinervitermes* spp. (Fig. 4.4.). However, *Trinervitermes* is such a reliable food resource that it has permitted the aardwolf to specialize on it as its major food, regardless of the possible disadvantages of doing so (Richardson 1987a).

4.6.2 Hunting and Foraging

Hyaenids show a wide variety of hunting and foraging specializations. Although the family consists of only four living species, they range from pack or clan hunters to solitary foragers. The large differences in diet between brown and spotted hyaenas are reflected in comparably large differences in foraging and feeding behaviour (Mills 1990).

Spotted Hyaenas. When hunting, the spotted hyaena does not show the stealth and agility of other large carnivores. Nor does it have great speed or determination as few spotted hyaenas attempt to attack vigorous and aggressive prey. They do have a remarkable tenacity, and an acute sense of smell and hearing. These characteristics are all employed when hunting (Deane 1962; Mills 1990). Hunting occurs mainly at night when their visual and olfactory advantages give them an edge over their prey (Bertram 1979). Spotted hyaenas can also follow a scent-trail, especially when the ground is damp (Mills 1990).

Whole clans, or smaller groups of spotted hyaenas, commonly hunt by long-distance chases. The prey is run down gradually until it is exhausted, just as African wild dogs do. In the Serengeti ecosystem, group hunting increases the chances of hunting success. In some instances the benefits gained from group hunting are probably the most important factor contributing to the development of social life in some carnivores (Bertram 1979), as in spotted hyaenas.

Not all hunting by spotted hyaenas in the Serengeti National Park is undertaken by groups or clans. Solitary hunting also occurs, but is directed mainly at smaller prey as in the Kruger National Park and the southern Kalahari (Kruuk 1966, 1972b; Bearder 1977; Mills 1990). When hunting as a group, spotted hyaenas will simply charge a herd of herbivores, each hyaena, or several together, biting at the legs and flanks of a fleeing individual until it is killed (Kruuk 1966, 1972b). Spotted hyaenas have also developed a range of behavioural, morphological and physiological adaptations which enable them to survive on highly variable migrating prey resources. These include long-distance foraging trips to the terrritories of other clans. This system of regular commuting to food resources is unique to spotted hyaenas in the Serengeti (Hofer and East 1993a-c). Although the spotted hyaena of the Savuti also commutes, it does so for water and not to forage (Sect. 4.5.1; Cooper 1990b).

As elsewhere, female spotted hyaenas in the Kruger National Park, have readier access to food because of their social status. The males compensate by moving greater distances when foraging, by consuming larger amounts of low-quality food, and by feeding more readily under adverse conditions (Henschel and Skinner 1987). In the Kruger National Park, spotted hyaenas usually forage in small groups, travelling up to 90 km between substantial meals as opposed to 33 km in the more arid southern Kalahari (Henschel and Skinner 1990). The reason for this is unclear as the clans in the Kruger National Park hunt cooperatively and efficiently (Bearder 1977).

Eloff (1964) first described the hunting method of the spotted hyaenas in the southern Kalahari. It is now generally known that spotted hyaenas there hunt and kill 55.3% of the carcasses on which they feed. However, the actual hunting frequency is probably higher. Spotted hyaenas seem to hunt as efficiently in the Kalahari as they do in East Africa, but they hunt more frequently than in the Kruger National Park where they scavenge more (Henschel 1986; Mills 1990). The mean distance at which prey is detected by hearing is 2.4 km in the southern Kalahari. Smell is the most important sense by which spotted hyaenas detect carrion. When using smell, the mean downwind distance at which they can detect a carcass in the southern Kalahari is 3.2 km (maximum: 4.2 km). In contrast, they cannot even find a strong-smelling carcass upwind. Live prey can also be detected when spotted hyaenas are downwind, but not surprisingly over shorter distances (mean: 1.1 km; maximum: 2.8 km) than carcasses. Sight is particularly important in hunting, and spotted hyaenas have excellent night vision. In the southern Kalahari, they detect potential prey visually in the open, flat, riverbed habitat in 60.3% of all prey contacts. When hunting gemsbok, the mean chase distance increases with increasing age of the prey up to subadult age, but decreases again for adults. These hyaenas clearly hunt their prey selectively according to species and age group. The prey is also selected by numerical abundance, the hyaena's hunting abilities and the prey's anti-predator behaviour (Mills 1990). In the Kalahari, there is no advantage to spotted hyaenas in foraging in large groups as this necessitates maintaining fairly large territories in a habitat in which food is scarce (Mills 1978a).

Although spotted hyaenas live in clans of up to 50 members in the Savuti Marsh, they normally forage in groups of one to four individuals. These hyaenas are particularly adept at kleptoparasitism, and often steal carcasses from lion prides lacking large males. The spotted hyaenas of the Savuti locate 78.5% of all lion kills within 30 min of the kill being made. Of all the lion prides found feeding on a kill, 44% have no adult male in attendance. Once such a pride is discovered on a kill by a pack of spotted hyaenas, it is mobbed when the hyaenas outnumber the lions by a least 2:1. The kill is then either eaten jointly by the lions and hyaenas, or it is taken over completely by the hyaenas when the lions are outnumbered by 4:1 or more. The presence of an adult male lion changes the balance of power drastically. When chased by an adult male lion, spotted hyaenas leave the vicinity of of the kill in 61% of cases (Cooper 1991).

Despite kleptoparasitism, the spotted hyaenas of the Savuti acquire 75% of their food through hunting, but most of the prey killed has suffered from a previous injury. They forage in small groups which increases their chances of locating the scattered herds upon which they prey. However, all clan members will congregate to feed on the carcass. These hyaenas mainly hunt at night. This is probably more to avoid the intense daytime heat than the need for darkness as hunting cover. On cool days, spotted hyaenas hunt in daytime too. The hunting technique involves approaching the prey in fan formation, usually downwind. An initial charge scatters the herd. The target animal is then selected visually. It is usually an individual that is vulnerable. Prey is only chased and killed if a suitable victim is detected. These hyaenas consequently have relatively low kill success (Cooper 1990a).

Clan size is relatively small in the spotted hyaenas of the Etosha National Park. The hyaenas there forage in groups of one to five. However, larger hunting groups are formed when hunting Burchell's zebra (Gasaway et al. 1991). In Etosha, spotted hyaenas kill 75% of their food. Further west in the more arid central Namib Desert spotted hyaenas only kill 50% of their food (Tilson and Henschel 1986).

Although the Moremi Wildlife Reserve is not an arid area, the spotted hyaenas there show an interesting approach to hunting. A single hyaena or a clan will pursue groups of lechwe *Kobus leche* into the water of the floodplains. Spotted hyaenas swim well, and the prey is caught and drowned. The carcass is then dragged to the water's edge where it is eaten. This hunting method is common (Child and Robbel 1975). It further underscores the flexibility of the hunting behaviour of spotted hyaenas.

Striped Hyaenas. Whereas striped hyaenas usually forage alone, they sometimes forage in groups of two or more in different locations (Rieger 1979a; Bouskila 1984). In the Negev Desert, one hyaena, which has recently found food, will return to the other members of the group and lead them to it (Bouskila 1984).

Brown Hyaenas. The largest difference behaviour between brown and spotted hyaenas is in their hunting. Hunting occupies only a small part of the brown hyaena's foraging budget, but a spotted hyaena spends much time and energy in doing so. When a brown hyaena does hunt, it does not demonstrate a well-structered strategy, but a primitive chase and grab. The brown hyaena forages alone, and investigates a wide variety of potential food. Brown hyaenas in the southern Kalahari are able to capitalize more efficiently on changes in food availability than are spotted hyaenas. Although brown hyaenas are solitary foragers, whereas spotted hyaenas frequently forage in groups, clan sizes do not differ in the southern Kalahari and a brown hyaena clan can well be larger than that of the spotted hyaena (Mills 1990).

Small mammals are hunted rarely by brown hyaenas in the southern Kalahari, and when they are, they are successful in only 4.7% of cases. Brown hyaenas never hunt large herbivores there. Of the carnivores, a single kill of a bat-eared fox has been recorded. Smaller birds such as the black korhaan *Afrotis afra* are occasionally snapped up and may be a more common prey of brown hyaenas in the southern Kalahari than has previously been thought (Mills 1990).

The only time when the brown hyaenas of the southern Kalahari appear to hunt purposefully is when they hunt springbok lambs. Even this is rare, however, and the success rate of such hunts is only 6% compared with 31% for spotted hyaenas (Mills 1990). Although fairly primitive, hunting by brown hyaenas is important as it provides highly nutritious food (Mills 1977; Owens and Owens 1978). Smell is the major sense used to detect the kills of other predators. Brown hyaenas can also follow scent trails, especially when the ground is damp. Their hearing is acute, and this enables them to find food they cannot smell. In addition, they have excellent night vision (Mills 1990). Carrion is located mostly by smell but the cries of an animal being killed by other predators are also used in the location of food (Mills 1977).

Hunting is more prevalent in brown hyaenas in the central Kalahari when carrion is scarce than at other times. The brown hyaenas there usually hunt alone, but socialize at carcasses. While foraging, a network of paths is used extensively both by spotted and by brown hyaenas. These paths are maintained through pasting. The paths connect carcasses up to three months old, and also lead to waterholes. The foraging distances per night in the dry season are 20 to 30 km per night, but in the wet season they are reduced to 10 to 20 km per night. During the rains, brown hyaenas forage where game and other predators congregate (Owens and Owens 1978). The short-tailed gerbil *Desmodillus auricularis* is probably hunted by brown hyaenas in the central Namib Desert (Skinner and Van Aarde 1981), but their main food source consists of dead seal pups as already said (Stuart and Shaughnessy 1984).

Aardwolves. The foraging behaviour of *Trinervitermes* spp. greatly influences foraging by the aardwolf. These termites cannot tolerate direct sunlight and regularly forage at night in large, concentrated groups on the surface (Kruuk 1972c; Richardson 1987b). The harvesting and storage capabilities of *Trinervitermes* spp. are not well developed. Consequently these termites have to forage more or less throughout the year, unlike those more efficient harvester termite species which show more seasonal surface activities (Kruuk 1972c). This behaviour is one of the vital forces which has helped to develop the aardwolf's specialization on *Trinervitermes* spp., and hence its adaptation to arid environments.

Foraging by aardwolves in the grasslands of the Serengeti National Park occurs at the slow speed of 2 to 3 km/h in an irregular path quite unrelated to wind direction. However, individual termite patches may be located upwind and then primarily by scent. Hearing is also employed at distances of 0.5 to 1 m commonly, and up to 2 m occasionally. As soon as it begins to rain, an aardwolf will stop foraging. The raindrops either blanket the sounds of foraging termites, or the termites move underground (Kruuk 1972c). The foraging speed of the aardwolf in the arid Northern Cape Province is determined by the number of termites available. The mean foraging speed, however, is 1 km/h when feeding, and 2.3 km/h when not feeding (Richardson 1987b).

Aardwolves forage singly and although ranges overlap with those of bat-eared foxes, interactions between these two species are rare as stated earlier (see Canidae; Bothma et al. 1984). Foraging involves patch utilization in the Namib Desert. Up to 15 min of feeding take place in a given patch. Food patches can be up to 100 m apart, and their richness is reflected in the time spent in each. The speed of foraging is 1 km/h in the Namib Desert as it is in the Serengeti and the Northern Cape Province. When foraging, an aardwolf in the Namib moves slowly and noiselessly forward, with the ears cocked and the head bent down. It follows an irregular route and licks up termites from the sandy substrate (Bothma and Nel 1980). The nutritional requirements of an aardwolf determine its foraging range. It is estimated that in the Northern Cape Province an aardwolf requires about 3000 termite mounds per territory to be able to survive (Richardson 1987c).

4.6.3 Killing

The canines of hyaenids other than the aardwolf, are specialized to cope with a large amount of bone. They are primarily adapted for killing by biting (Anyonge 1996).

Spotted Hyaenas. In the Serengeti, group hunting by spotted hyaenas increases the kill rate. A solitary spotted hyaena hunting wildebeest calves has a kill rate of 15% of all animals hunted as opposed to 23% when two hyaenas hunt together, and 31% when the hunting group numbers three or more (Kruuk 1972b). The spotted hyaena apparently does not use any special techniques to kill adult gemsbok in the southern Kalahari as the lion does (see Felidae; Eloff 1964, 1973a,c, 1984; Mills 1990). That the gemsbok can successfully defend itself when hunted by spotted hyaenas is clear from the low kill rate of 14% of adults (Mills 1990).

A crucial factor contributing to the success of a gemsbok calf hunt by spotted hyaenas is whether the calf can be separated from the herd. There is, however, no relationship either between the kill rate of gemsbok calves and the size of the herd or the group size of the hyaenas. There is also no indication that a solitary hyaena is less successful than a group when hunting gemsbok calves. When several spotted hyaenas encounter a gemsbok herd, they appear to hunt independently until a calf has been isolated and the chase begins. When hunting adult gemsbok, however, a group of four hyaenas is the smallest which can be successful. Based upon the marrow fat of the long bones, it appears that gemsbok in poor condition are selected. Of all gemsbok hunts, only 7.6% end in a kill (Mills 1990).

The kill rate of blue wildebeest of all ages by spotted hyaenas in the southern Kalahari is 39%. This is close to the 34% kill rate for wildebeest by the same carnivore in East Africa. There is no obvious relationship in the Kalahari between the chase distance by spotted hyaenas and the age of wildebeest. However, the mean chase distance of 1.0 km leading to a kill is greater than the 2.1 km in unsuccessful hunts. Overall, the wildebeest of the southern Kalahari and East Africa are under similar hunting pressure from spotted hyaenas, but in the Kalahari there is less pressure on calves (Mills 1990).

The kill rate of springbok lambs by spotted hyaenas in the Kalahari is 31%. Although the most common ungulate in the dry riverbeds which are favoured by spotted hyaenas, springbok are largely ignored, except in summer during the peak lambing season. Chase distances for springbok lambs are short, and chases are soon given up. The mean chase distance is only 644 m (maximum: 1700 m) in unsuccessful hunts. Other game are usually chased for longer distances. Springbok are mainly hunted by solitary hyaenas (Mills 1990).

Other ungulates are hunted less frequently than gemsbok and blue wildebeest in the southern Kalahari. Eland are only available sporadically, but when they are present, their calves are often hunted. Only one kill of a subadult eland is known. In that case three spotted hyaenas were involved. It occurred after the hyaenas had followed the eland herd for 23.8 km. They killed a subadult bull after a struggle stretching over 1 km. The fleet-footed red hartebeest, present in small numbers in the Nossob riverbed, is rarely hunted. The small, but equally fleet-footed steenbok, is sometimes hunted by solitary spotted hyaenas. Their 33% kill rate (Mills 1990) is much higher than that of leopards living in the interior dunes, as stated earlier (Bothma et al. 1997).

Porcupines are occasionally encountered by spotted hyaenas and may sometimes be killed. Cape hares are also killed occasionally. The overall killing success of spotted hyaenas in the southern Kalahari varies between clans. It does not appear to be related to differences in hunting ability, but rather to prey selection. Less successful clans particularly select prey which are more difficult to kill (Mills 1990). In the Savuti, killing success is low (Cooper 1990a). Surplus killing occurs among spotted hyaenas in the Serengeti (Kruuk 1972a). It must be less frequent, if found at all, in arid regions where prey are not usually present in large numbers.

Other Hyaenids. As striped hyaenas are primarily scavengers, their killing technique is little known. A striped hyaena will run and jump into the air to snap up a flying insect. Brown hyaenas, on the other hand, do not jump into the air when catching insects. In the southern Kalahari, brown hyaenas are quite adept as occasional hunters, killing 6% of the biomass of their food. Some brown hyaenas may even become habitual killers. One female learned to hunt black korhaans effectively (Mills 1990). Aardwolves simply lick up their food from the soil surface.

4.6.4 Feeding

Spotted Hyaenas. As often as not a spotted hyaena manipulates its food with its front paws. Spotted hyaenas generally do not chew their food as much as dogs do (Van Valkenburgh 1996). Their digestive abilities, however, allow them to digest carcasses more efficiently than cats can. Because of competition, spotted hyaenas feed as rapidly as possible at a large carcass. There is, however, little squabbling among clan members when feeding together (Bertram 1979), even when up to 30 hyaenas feed at the same carcass (Mills 1990). In areas of high hyaena density, the carcass of any animal smaller than a buffalo is usually reduced to a skeleton within 30 min (Frank 1986b). In an arid region such as the Namib Desert, however, where spotted hyaenas often feed alone, or the feeding group at most consists of a female and her young, large carcasses are consumed leisurely over several nights (Mills 1990).

A spotted hyaena can easily consume up to 30% of its body mass in food in a single meal (Frank 1986b). Sodium and water turnover rates have been used in the case of free-living spotted hyaenas to derive food consumption rates based upon estimates of sodium influx. Such estimates indicate a consumption rate of 3.8 kg of food per hyaena per day compared with 6 kg per day by a lion (Green et al. 1984). Spotted hyaenas in the Kruger National Park indeed consume about 3.8 kg of food per individual per day (Henschel and Skinner 1990).

At permanent den sites, such as granite or calcrete caves in the Kruger National Park, spotted hyaenas accumulate considerable quantities of bone. Fewer bones accumulate at temporary dens such as aardvark burrows or culverts underneath roads (Skinner et al. 1986). Spotted hyaenas also collect bones around their dens in the southern Kalahari (Mills and Mills 1977). Of carcasses carried to dens, 12.7% are taken further than 1 km. Moving food away from its source results in reduced intra-group competition when feeding (Skinner et al. 1986).

Female spotted hyaenas in the Kruger National Park have priority at carcasses. Hence they have greater food intake than males (Henschel and Skinner 1990). Scavenged food is eaten slower than killed food in the southern Kalahari. If a carcass is not devoured completely in one night, the hyaenas usually drag it under a bush and spend the day resting close by. Adult females with cubs at the den, however, will often abandon half-eaten carcasses to return to their cubs (Mills 1990).

Once a kill has been made in the Savuti, clan members converge on it (Cooper 1990a). In the central Namib Desert, where spotted hyaena densities are low, carcasses are often left unattended during the day (Mills 1990). Nevertheless, these hyaenas may form groups larger than three at carcasses (Tilson and Henschel 1986). Along the Kuiseb River Canyon of the Namib Desert, each adult clan member eats 8.7 kg of food per night from large, fleshy ungulate carcasses. This is considerably more than the 3.8 kg per night generally expected for spotted hyaenas, and found in the Kruger National Park (Henschel and Tilson 1988; Henschel and Skinner 1990; Green et al.1984). In the Namib, the lowest-ranking male gets considerably less than the other clan members to eat. These males only eat about 4.35 kg per night. Antelopes of medium size are devoured rapidly. As the size of the group which feeds on smaller carcasses is usually smaller than on large ones, it is possible that each individual hyaena gets more food per night when feeding on carcasses of medium than larger size. The lower rate of food consumption found elsewhere, such as in East Africa, when compared with the Namib Desert possibly relates to clan size, territory size, density and quality of the food, activity periods and distances moved daily. Adult spotted hyaenas in East Africa are also 20% smaller than those of southern Africa (Henschel and Tilson 1988).

Spotted hyaenas rarely store food in caches. When this does occur, spotted hyaenas in East Africa store food at waterholes. However, in the southern Kalahari, food is stored in tall grass and under bushes as well as around waterholes (Mills 1990). Food storing by spotted hyaenas has not yet been observed in the arid bushveld of the Timbavati Private Nature Reserve (Bearder 1977). In the Kruger National Park food is cached by spotted hyaenas at different sites, usually in grass clumps, under bushes or even under water (Skinner et al. 1986). Spotted and brown hyaenas usually return within 24 h to retrieve stored food (Mills 1990). Spotted hyaenas do not appear to have preferential feeding sites. They normally tear a carcass apart as quickly as possible (Skinner and Ilani 1979).

Striped Hyaenas. The striped hyaena begins to feed at the anus, and the buttocks are the preferred food (Skinner and Ilani 1979). Food storing is known in striped hyaenas (Mills 1990), but no other details are known. In the Judean and Negev Deserts, the feeding sites of striped hyaenas are dependent upon waterhole distribution. In the Serengeti, striped hyaenas cache food. A supply of cached food is maintained upon which a striped hyaena can draw in times when food is scarce (Mills 1990).

Brown Hyaenas. In the southern Kalahari, brown hyaenas usually eat alone when their food consists of small pieces of bone, old leg remains and skulls, which is usually the case. Smaller carcasses are consumed more easily than larger ones because brown hyaenas normally struggle to break off and consume the larger bones. Several brown hyaenas may gather at a single carcass, and in the southern Kalahari there is little evidence of the strict dominance hierarchy which exists at a carcass in the central Kalahari (Owens and Owens 1978, 1979b; Mills 1990). In the latter region, the viscera of fresh carcasses are eaten first. The carnassial teeth are then used to cut the skin before the meat is eaten, and to lever the femur from its socket. The forelegs of the carcass are separated in the same way. The brown hyaena feeds in a more leisurely manner, but eats considerably less (4.5 to 5 kg of fresh meat and 1.5 to 3 kg of leg or bone and skin per individual per sitting) than do the spotted hyaenas. At 58.6% of all carcasses, two or more brown hyaenas feed together, socializing at the feeding site. When at a den in the central Kalahari they may feed inside or outside, but there is one area in which most of the old bones and other remains accumulate (Owens and Owens 1979b).

Like the striped hyaena, the brown hyaena in the southern Kalahari stores excess food for later use, but usually recovers it within 24 h. Ostrich eggs are favourite items for storage, and whole nests containing as many as 26 eggs may be destroyed in this way. A brown hyaena can easily carry an ostrich egg in its mouth. In the southern Kalahari, the maximum distance over which eggs are moved is 6.8 km, but the distance normally varies from 150 to 600 m. When a brown hyaena finds a large amount of food, it will quickly eat some of it, carry some away to cache, and then will return to devour the rest. A second brown hyaena will sometimes return with the first to eat the stored food. Ostrich eggs, bones and old pieces of hide may be stored for long periods (Mills 1990). Of all carcasses eaten by brown hyaenas in the central Kalahari, 70% have parts which are cached. Some food is carried 2 km or more toward the den. In some cases also, the cache of one brown hyaena may be discovered and eaten by another one (Owens and Owens 1978). Brown hyaenas also cache food for later use in the Namib Desert (Mills 1990).

Aardwolves. Having discovered a foraging colony of termites, the aardwolf licks them up with its broad tongue (Fig. 4.5). In feeding, large amounts of sand and other debris are also ingested. The proportion of soldiers eaten increases as time passes at a given food patch, because soldiers emerge when workers are disturbed. As already mentioned, the soldiers secrete various noxious and probably distasteful terpenoids (Kruuk 1972c; Anderson et al. 1992; Richardson and Levitan 1994). It is possible that once these secretions exceed a specific threshold of tolerance, the aardwolf may stop feeding at that patch and move to another undisturbed one.

The aardwolf is specialized for utilizing a limited food resource. The large quantities of sand and debris ingested with its food makes molars redundant, and they are extremely reduced in an aardwolf. In contrast, all other hyaenids have a wide dietary spectrum and well-developed molars. Most conspicuous is the difference between the aardwolf and its look-alike, the striped hyaena (Kruuk 1972c). The aardwolf does not dig for its prey as do other myrmecophores. Nevertheless, it does have a number of anatomical features which are clearly adapted to its diet. These include a broad, spatulate tongue covered with large and hardened papillae, a broad palate, large mandibular salivary

Fig. 4.5. A foraging aardwolf in the central Namib Desert

glands, and a muscular pyloric region of the stomach. The relatively short small intestine facilitates the rapid processing of large volumes of termites, while the distensible large intestine allows for the accumulation of much faecal material (Anderson et al. 1992). The exceptionally large salivary glands suggests that the saliva of the aardwolf may contain properties for neutralizing or detoxifying terpenes. However, the fact that the faeces smell strongly of terpenes, suggests neutralization rather than detoxification (Richardson and Levitan 1994). The negative aspect of the specialization in diet is that the aardwolf has become so adapted to in licking up small anthropods that, combined with the almost complete degeneration of the molars, it is generally unable to handle larger food items even in times of food shortage. Moreover, the aardwolf cannot now inhabit areas where its main food resource is absent (Richardson 1987b).

4.6.5 Defecation

Spotted hyaenas are usually believed to defecate in characteristic latrines near prominent landmarks. These serve a scent-marking function as said earlier (Mills and Gorman 1987). However, the spotted hyaenas of the Savuti seldom use latrines (Cooper 1990b), but the reason for this is unknown. Deposition of faeces in latrines by the striped hyaena also has a scent-marking function (Kruuk 1976), but little more is known about its defecation habits. Brown hyaenas likewise deposit faeces in latrines which have a scent-marking function. However, whereas brown hyaenas deposit all their faeces in small and clearly demarcated areas, those of spotted hyaenas may cover some 100 m^2 and may be deposited both by members of the clan or by solitary individuals (Skinner and Van Aarde 1981).

The defecation habits of the aardwolf are unique and characteristic. The aardwolf defecates in any of a number of faecal middens which are distributed throughout its range. As many as four active, and many more inactive middens may occur in a territory. A midden is a clearly defined circular to oval area, 1 to 2 m in diameter which is thickly carpeted with old droppings (Kruuk 1972c). The faeces of the aardwolf, as already mentioned, has a characteristic terpene smell derived from the secretions of the soldier termites upon which it feeds (Richardson 1990).

Large defecations are deposited in these middens, and are then buried by sand. The faeces are also often urinated on after burial. This is not necessarily linked to scent-marking, but may play a role in mating. In summer, an aardwolf in the Northern Cape Province defecates three to six times per night in various middens. Hence these middens are focal points of aardwolf activity. When a midden occurs on a territory boundary, both neighbours scent-mark it by pasting, often covering each other's scent-marks (Richardson 1990).

The exact biological role of the faecal midden of the aardwolf remains unclear (Kruuk 1972c). If a midden has a territorial function, then it seems illogical to bury the faeces under sand, as this must reduce its signalling value. Aardwolves probably detect their prey at low concentrations by the smell of the terpenes secreted by them. As the faeces also smell strongly of terpenes, their occurrence might cause confusion. By burying the faeces, this confusion may be reduced (Nel and Bothma 1983; Richardson 1990).

Faeces are buried only by adult aardwolves in the central Namib Desert. Younger aardwolves do not bury their faeces. Several adults will use the same midden in a given night (Nel and Bothma 1983), especially if the midden is on the territory boundary as has been found to be the case in the Northern Cape Province (Richardson 1990). In the Namib Desert, the middens of aardwolves are not defecated on by ungulates. However, this does happen in East Africa, the southern Kalahari and elsewhere in southern Africa. Increased aridity may play a role here. The density of middens in the Namib Desert is 0.38 middens per km^2, and the nearest neighbouring midden is a mean 237 m away (Nel and Bothma 1983). The middens of the aardwolf in the Namib Desert are oval, with mean lengths of 5.7 m and widths of 4.5 m. These middens are larger than in East Africa where they are roughly circular with a mean diameter of 1 m (Kruuk 1972c; Nel and Bothma 1983). The middens in the Namib Desert may somehow be correlated with the foraging pattern of the aardwolves. When the need to defecate arises, an aardwolf knows exactly where the nearest midden is (Nel and Bothma 1983).

Defecation by aardwolves in the Namib Desert takes place at any time during the night, but mostly on emergence from the burrow. Urination usually follows, but may also occur at any time or place when foraging. Urination spots on open ground where defecation is not involved are covered by scraping sand over them. The consistent burial of faeces under sand in arid areas, such as the Namib Desert and the southern Kalahari, and the perfunctory occurrence of this in more mesic areas in South Africa, is perhaps related to environmental features. In arid, open areas, wind will disperse the odours emanating from a midden more readily than in areas which are more densily vegetated (Nel and Bothma 1983; Richardson 1990).

The faeces of an aardwolf provides a rare source of moist, organic matter in an arid region. This resource is used opportunistically by small invertebrates which may become dependent upon the aardwolf's presence. Within two minutes of deposition, the fresh faeces of an aardwolf in the Namib will already have attracted small beetles of the families Histeridae and Scarabaeidae. The faeces also soon contains numerous live Diptera (Drosophilidae) larvae, which are buried alive with the faeces, and pupate within five hours of deposition. The fresh, moisture-laden dung is used as a food resource for beetles, while Histeridae prey on Diptera larvae. Scarabaeidae beetles are common in the dung of mammals throughout Africa (Bothma and Nel 1983).

4.6.6
Water Use

The presence of water is essential to spotted hyaenas (Fig. 4.6), and they cannot inhabit arid regions without it (Skinner and Van Aarde 1981). Although they do not normally derive a significant fraction of their water intake from drinking (Green et al. 1984), spotted hyaenas in arid regions need it when their food resource becomes unreliable. In the southern Kalahari, spotted hyaenas supplement their water balance more readily from their food than brown hyaenas, because they kill game more frequently. Both species of hyaena also eat wild fruits which are high in moisture content. The tsama melon, for instance, contains up to 90% water. Spotted hyaenas are also mainly nocturnal, which helps them to conserve water (Mills 1990).

As stated earlier, commuting into the territories of neighbours to find water is unique among the spotted hyaenas of Savuti. This happens especially during droughts, and the hyaenas always return to their own territories after drinking. In years of sufficient rainfall, every spotted hyaena clan of the Savuti has its own source of near-permanent water within its territory (Cooper 1990b). The spotted hyaenas of the Etosha National Park also depend upon permanent waterholes in their territory. They also use the body fluids of their prey to maintain large foraging areas, even when water is scarce (Gasaway et al. 1991). The spotted hyaenas of the central Namib Desert are concentrated along the Kuiseb River Canyon. Although the river is ephem-

Fig. 4.6 Spotted hyaenas are dependent upon a reliable source of water within their range

eral, this canyon contains small, isolated pools of water for most of the year. South of it stretches a vast, waterless sand desert. In years of drought, the ungulates of this desert also congregate along the canyon, and their body fluids supplement the water supply of the spotted hyaenas (Tilson and Henschel 1986).

For the striped hyaena to survive, water must be available within a radius of 10 km. For this reason, striped hyaenas are absent from most of the Sahara and central Arabian deserts (Rieger 1979a). Striped hyaenas readily drink water when available, as it is in the Serengeti, but in the Negev Desert they go without drinking for long periods (Kruuk 1976).

The brown hyaena drinks water when it is available, but is independent of it. It is well-adapted to live in arid environments such as the Namib Desert where there is little or no free water (Eaton 1976a). Its diet provides less moisture than that of the spotted hyaena. In the southern Kalahari, most of the brown hyaena's vertebrate food is already dry when eaten, and wild fruits are used extensively to supplement its water intake (Mills 1990). The wild fruits mostly eaten are the tsama melon and the gemsbok cucumber (Mills 1978a). In the central Kalahari, wild fruits are an equally important source of moisture for brown hyaenas during the dry season. By volume, some 50% of the faeces collected then may contain the rinds or seeds of fruits. When water is available, however, the brown hyaena drinks at least once a night, and often two to three times while foraging (Owens and Owens 1978).

In being independent of free water, the brown hyaena of the Namib Desert is well-adapted to conserve moisture. It thrives in the coastal region where food is abundant. Nevertheless, severe extremes in temperature also occur. Bitterly cold south-west winds sweep along this coastal belt. Its long hair and pilo-erection helps the brown hyaena to adjust to such adverse climatic conditions. The independence from water of the brown hyaena, and the thermal insulation provided by its long fur (mean: 112 mm) compared with the short fur (mean: 33.4 mm) and greater water dependence of a spotted hyaena, enable brown, but not spotted hyaenas to live along the Namib Desert coast (Skinner et al. 1984).

Aardwolves are independent of water as they get all their moisture from the termites eaten. During long, cold spells, however, when termites are relatively inactive, they may walk long distances to drink. This is the only time when they do so (Richardson 1985).

4.7
Population Ecology

4.7.1
Spotted Hyaenas

The resident prey population in the Serengeti ecosystem does not limit the spotted hyaena population as suggested by earlier authors. The two factors which are currrently considered to be most important in affecting the Serengeti spotted hyaena population are the removal of 8% of breeding females per year by snaring and poisoning by hunters, and hyaena social factors such as infanticide and siblicide (Hofer and East 1995).

From 1969 to 1986, there was an enormous increase in the number of wildebeest in the Serengeti National Park. In response, the spotted hyaena population almost doubled in that period. This was probably because the length of the commuting trips of the females was reduced. In turn, this has led to improved survival of the young (Hofer and East 1995). Despite the fact that their abundance does not regulate the spotted hyaena population, nearly all the wildebeest which die in the Ngorongoro Crater are killed by these hyaenas. Furthermore, the wildebeest population of this crater is fairly stable. In the Serengeti, many wildebeest die from causes other than predation by spotted hyaenas. Their population cannot therefore be regulated by spotted hyaenas. These hyaenas may influence the age-structure and turnover-rate of their prey populations, but do they not determine the mean levels of these populations. In both areas, the numbers of spotted hyaenas seem to be linked to food, but in different ways. In the Serengeti the food supply of small hyaena cubs seems to be a regulating factor. In the Ngorongoro Crater, on the other hand, it is the food supply of adult hyaenas that may regulate the spotted hyaena population (Kruuk 1970).

The number of surviving young of spotted hyaenas in the Masai Mara National Reserve is strongly affected by the rank of the females. The highest ranking females produce 2.5 times more surviving young of both sexes than lower ranking females (Frank et al. 1995). The maximum life-span of a spotted hyaena in captivity is between 18 and 25 years (Van Jaarsveld et al. 1987). In the Masai Mara area, females of higher rank begin breeding at a younger age than those of lower rank. They produce living young earlier, and their life expectancy is higher. The number of young which are present in the clan increases during the life-span of such a female. It is also possible that at low densities of hyaenas in otherwise favourable areas, the females may adjust the sex ratio of their young by producing more female than male cubs (Frank et al. 1995). In the Serengeti, most spotted hyaena cubs die of predation from lions or of siblicide (Brocklesby and Vidler 1965; East et al. 1989). Werdelin and Solounias (1991) also regard siblicide to be a significant cause of death in spotted hyaena cubs.

Spotted hyaena cub mortality in the southern Kalahari is low, but proportionally more spotted hyaenas die in their prime there than elsewhere. The spotted hyaena population of the southern Kalahari is thus relatively young. Lion predation is one cause of death of spotted hyaenas in the Kalahari, as it also is in East Africa where it is the major cause of death. However, in the southern Kalahari the main cause of death (43% of all known cases) is rabies. With the relatively small size of most Kalahari clans, however, this disease usually spreads only within and not often between clans (Mills1990).

In the southern Kalahari, predation by spotted hyaenas appears to have little influence on the prey population (Mills1990). In contrast, preliminary evidence suggests that predation in Savuti probably affects not only the resident impala population, but also through predation on foals, the migratory zebra population. At high population densities, the spotted hyaenas of Savuti have the potential to regulate the populations of those prey that do not migrate (Cooper 1990a).

In the central Namib Desert, the spotted hyaena population will eat some 4.8% of the mean ungulate biomass, removing 14.3% of the gemsbok population and 2.2% of the Hartmann mountain zebra population in a year. However, this rate of depletion does not appear to limit the populations of either of these prey, even though the spotted hyaena is the most numerous large carnivore present (Skinner and Van Aarde 1981; Henschel and Tilson 1988). Predation by spotted hyaenas is, however, not merely additional to other sources of mortality, but also interacts closely with them. When alternate causes of death, such as starvation, are numerically important, predation by spotted hyaenas becomes insignificant because there are many fresh carcasses available for scavenging. However, when other causes of death are largely absent, the spotted hyaenas become independent mortality agents (Tilson et al. 1980). The population density of spotted hyaenas in the Namib Desert therefore appears to be regulated by a combination of factors, including resource availability, which in turn affect the recruitment rate, intensity of competition, local dispersion and the aggression of clan members (Tilson and Henschel 1986).

4.7.2 Other Hyaenids

In contrast with spotted hyaenas, the brown hyaenas of the southern Kalahari do not suffer from rabies. This is probably because they are largely solitary by nature (Mills 1990). The most important variable in the regulation of the brown hyaena population there is subadult emigration (Mills 1983). Both sexes emigrate from their natal clans. Most, if not all, males eventually leave their natal clan, but a few females may remain with it for life. Immigration is confined to males, and is not an important agent of population regulation. Cub mortality is low and most adults live to old age. Violent death resulting from fights is the most common cause of death. Tooth-wear may also lead to starvation and death in old age (Mills 1990). In the central Kalahari, cub mortality appears to be higher than in the southern Kalahari. One in every six litters in the central Kalahari fail to survive (D.D. Owens and Owens 1984). Aardwolf cubs are also known to die from food shortage in winter (Richardson 1987b).

4.8 Interaction with Other Carnivores

The differences in the sizes of predator species, and their selection of habitat and hunting behaviour result in relatively little overlap in diet. The degree of exploitative competition is therefore slight, and the major factor in competition is interference. For example, the spotted hyaenas of the Serengeti kill many wildebeest calves, but do so when the herds are in the open and mostly beyond the reach of lions. Nevertheless, they not infrequently drive African wild dogs away from their kills. Exploitation competition is small on account of the numbers of prey available. All the Serengeti predators combined do not regulate the available prey base. This is because 80% of the prey resource is migratory while the predators tend to be more sedentary (Bertram 1979).

4.8.1 Spotted Hyaenas

There is more competition over food between spotted hyaenas in the Ngorongoro Crater than in the Serengeti. Adult hyaenas in the Crater also die at a younger age than in the Serengeti (Kruuk 1970). Amongst the larger carnivores of the arid bushveld of the Timbavati Private Nature Reserve, the spotted hyaena interacts most frequently with the black-backed jackal, lion and leopard. Cheetahs and African wild dogs are rare in the area (Bearder 1977). In the more mesic Umfolozi Game Reserve, however, lions, leopards, wild dogs and cheetahs all compete with spotted hyaenas for food (Skinner et al. 1992).

The spotted hyaena is ecologically less successful than the brown hyaena in the southern Kalahari (Mills 1989). The large predators present there can be divided into two groups on the basis of prey selection. Lions and spotted hyaenas prey predominantly on large to medium-sized mammals weighing more than 12 kg, and nearly 70% of their combined food consists of gemsbok and blue wildebeest. Leopards and cheetahs, on the other hand, hunt mainly smaller prey, while brown hyaenas seldom hunt at all. Competition for gemsbok and blue wildebeest by lions and spotted hyaenas is reduced by the fact that the lions mainly kill subadult and adult gemsbok and blue wildebeest, whereas spotted hyaenas kill calves, especially those of the gemsbok. Spotted hyaenas also scavenge more than lions do. There is direct competition, however, between spotted and brown hyaenas in scavenging medium to large-sized mammals (Mills 1990).

By killing adult gemsbok in the southern Kalahari, lions may also be depriving the spotted hyaenas there of an important source of food. In the central Namib Desert, where there are no lions, spotted hyaenas regularly kill adult gemsbok (Tilson et al. 1980; Mills 1990). In East Africa and in Savuti, lions and spotted hyaenas steal each other's prey (Mills 1990; Cooper 1991). In the southern Kalahari, however, the spotted hyaena usually waits for lions to leave a kill before moving in, and vice versa. Clashes between these two large predators are therefore rare, and neither has a clear advantage over the other. A group of spotted hyaenas may sometimes attack lions, even when food is not at stake. This particularly happens if the lion pride involved contains no adult males as said earlier. The lions usually win such clashes, but may occasionally jump into a tree to escape. Solitary lions may not be able to compete with a group of spotted hyaenas (Mills 1990).

Spotted hyaenas and leopards likewise suffer little from exploitative or interference competition for food in the southern Kalahari. Defence of prey depends upon the size and number of hyaenas present. Leopards are solitary and are usually dominated by two or more spotted hyaenas. However, the leopard can escape competition by taking its food up a tree. A group of spotted hyaenas will usually harass a leopard until it takes refuge in a tree, or down a suitable burrow. This may happen even in the absence of food. Although leopards are dominated in this way by spotted hyaenas, they seldom lose much food to them, or suffer in any other way (Mills 1990).

The prey of cheetahs and spotted hyaenas in the southern Kalahari shows little overlap. Hyaenas consequently gain little food from cheetahs. Spotted hyaenas occasionally chase cheetahs off a kill, usually at dusk when they pick up the sounds and smell of the hunt (Mills 1990). Overlap in prey selection between black-backed jackals and spotted hyaenas is small in the southern Kalahari. When it occurs, it is usually for springbok lambs. Black-backed jackals occasionally deprive spotted hyaenas of potential food, or scavenge from hyaena kills (Mills 1990).

Spotted hyaenas and caracal do not interact (Mills 1990). In the central Namib Desert, the spotted hyaena is the most abundant of the larger predators. Nevertheless, there is a clear geographical separation between the ranges of the spotted hyaenas that are found along the Kuiseb River Canyon, and the brown hyaenas which inhabit the Skeleton Coast. Despite the abundant food along the coast, the spotted hyaenas do not inhabit that area because of the cold and windy conditions which prevail there, even in summer (Skinner and Van Aarde 1981; Skinner et al. 1984).

As mentioned before, large groups of spotted hyaenas in Savuti often displace lion prides where there is no adult male in attendance. In return, 25% of all spotted hyaena kills are lost to adult male lions (Cooper 1990a). Having evolved in competition with spotted hyaenas, whose diet and ecological range significantly overlap theirs, group hunting has been developed by lions as a strategy that counters this competition. In Savuti, the spotted hyaenas will never challenge a group of lions which contains an adult male. When chased by an adult male lion, spotted hyaenas leave a lion pride's kill on 61% of all occasions (Cooper 1991).

Except when there are localized epidemics of disease amongst ungulates, competition for fresh carrion is intense among large carnivores in the Etosha National Park. This includes competition by spotted hyaenas with vultures, black-backed jackals and lions, all of which are day-active. This competition often leaves little for the nocturnal hyaenas to scavenge (Gasaway et al. 1991). On the other hand, the hyaenas' own nocturnal kills are relatively safe from interference by diurnal scavengers (Mills 1989).

4.8.2 Striped Hyaenas

The striped hyaena's major threat in Asia comes from man. Almost every part of the animal's body is used in some superstition or other, as a medicine or an aphrodisiac. In ancient Egypt, striped hyaenas were caught, tamed and used for hunting. In Israel the striped hyaena has adapted over many centuries to live close to man, with whom there is a minimum of conflict (Skinner and Ilani 1979). The diets of both striped and spotted hyaenas have several items in common. In the relatively small areas where their ranges overlap, this leads to competition. However, there is normally little contact between brown and striped hyaenas, especially in East Africa. Striped hyaenas stay away from other large predators, while smaller carnivores are either ignored or preyed upon (Kruuk 1976). In the Judean and Negev Deserts, the striped hyaena is one of the largest remaining carnivores. It is mainly a scavenger and has few or no competitors there (Skinner and Ilani 1979).

4.8.3 Brown Hyaenas

Except in the use of carrion, competition between brown and spotted hyaenas in South Africa is low. In the Kruger National Park, the brown hyaena overlaps with the cheetah in 72% of its prey species, with the leopard in 32%, with the African wild dog in 21%, with the spotted hyaena in 32% and with the lion in 29%. The extent of this competition is sufficient to be a limiting influence on the brown hyaena population there. Interspecific competition for carcasses is especially severe. Where spotted and brown hyaenas are sympatric in southern Africa, the spotted hyaena is more common (Eaton 1976a), except in arid regions such as the southern Kalahari (Mills 1990). In sympatric areas, the brown hyaena ranks low in predator hierarchy, dominating only the cheetah, and often only when a cheetah is alone (Eaton 1976a; Mills 1990).

Lion kills are an important source of food for brown hyaenas in the southern Kalahari, but are only scavenged after the lions have left. On rare occasions, lions will chase brown hyaenas away from the carcasses of animals that have died of causes other than predation (Mills 1990). Lions will also occasionally attack, maim and kill brown hyaenas for no apparent reason as they are not eaten. Such incidents seldom take place at carrion, but at the hyaena's den. Sometimes the brown hyaena is only mutilated and left to die a painful death (Eloff 1984; Mills 1990). Overall, however, the brown hyaenas of the southern Kalahari gain more than they lose from the relatively small lion population (Mills1990).

Leopards have little influence on brown hyaenas in the southern Kalahari, and vice versa. When a leopard takes its prey up a tree, it is not available to a scavenging hyaena. When a brown hyaena and a leopard meet while foraging, they keep their distance and show little interest in each other (Mills1990). However, in the central Kalahari, a brown hyaena occasionally appropriates a leopard kill (Owens and Owens 1978).

Based upon circumstantial evidence, the cheetahs of the dry Auob riverbed in the Kalahari Gemsbok National Park appear to provide a substantial amount of food for brown hyaenas. Carcasses of springbok killed by cheetahs during the day, and not fully consumed before nightfall, are usually eaten by brown hyaenas during the night. Along the dry bed of the Nossob river, the supply of carcasses by cheetahs is more sporadic as cheetahs are less abundant there. When scavenging a cheetah's food, the brown hyaena usually waits until the cheetah or cheetahs have left. Consequently there is little interaction between them, except occasionally when a brown hyaena may chase a solitary cheetah away from a kill made close to nightfall (Mills 1990).

Caracal are rare in the southern Kalahari, but brown hyaenas have been known to appropriate caracal kills. Merely moving towards the feeding caracal is enough to make it withdraw without resistance. The black-backed jackal is abundant in the southern Kalahari, and its diet includes scavenged food. It therefore does compete with the brown hyaena. However, a brown hyaena will often go towards a black-backed jackal if it can gain food by doing so. It will also devour a black-backed jackal carcass. Although black-backed jackals may deprive brown hyaenas of a substantial amount of food, these jackals cannot eat large bones. Brown hyaenas and black-backed jackals often feed together on a carcass, and the jackals also pick up left-over scraps after brown hyaenas have fed. Thus there is some degree of competition, but both gain and lose. As there are many more jackals than brown hyaenas in the southern Kalahari, however, the brown hyaenas probably lose more (Mills 1990).

Brown hyaenas tend to avoid areas of high spotted hyaena density. As the spotted hyaenas in the Kalahari Gemsbok National Park prefer dry riverbeds to other habitats, 93% of all brown hyaena dens are found in the dunes, while 69% of spotted hyaena dens occur in the riverbeds (Mills and Mills 1982). The almost exclusively nocturnal habits of brown hyaenas may also have evolved in response to competition from dominant diurnal scavengers, including vultures (Mills 1977).

Lions are known to kill brown hyaenas occasionally in the central Kalahari (Owens and Owens 1979a). Brown hyaenas avoid lion kills and never come closer than 200 m until after the lions have left. They may then approach the carcass cautiously. A brown hyaena will occasionally steal a leopard's kill, and readily appropriates the kills of cheetahs. African wild dogs do not tolerate brown hyaenas at their kills and in turn appropriate kills taken by brown hyaenas from cheetahs. Brown hyaenas leave spotted hyaenas severely alone when the latter are at a carcass, but they react strongly to black-backed jackals. These, in turn, often follow foraging brown hyaens (Owens and Owens 1978). Along the Skeleton Coast there also is a close association between brown hyaenas, black-backed jackals and the colonies of the Cape fur seal which provide both of them with their major source of food (Stuart and Shaughnessy 1984).

4.8.4 Aardwolves

The aardwolf exploits a resource neglected by most other mammals in Africa (Richardson 1987a,b). In the central Namib Desert, however, the foraging areas used by aardwolves overlap those of bat-eared foxes and black-backed jackals, both of which also eat insects. Therefore there is broad-based competition between the first two carnivores. The Cape fox also eats insects, but has a wide and varied diet and does not compete with either the aardwolf or the bat-eared fox. Despite their broad dietary overlap, however, there is a separation between the aardwolf and the bat-eared fox with respect to the type of insect mostly eaten (Bothma et al. 1984).

Neither the aardwolf nor the bat-eared fox increase the variety in their diet with increasing drought in the central Namib Desert. Therefore the level of competition does not increase, and there must be sufficient food for both. Different times of emergence and onset of foraging also limit competition, even when the same foraging range is used. The aardwolf and bat-eared fox favour short grass patches, a habitat rich in termites but which is avoided by the Cape fox. Black-backed jackals spend little time in any given food patch. Sufficient separation therefore exists between the aardwolf, bat-eared fox, Cape fox and black-backed jackal in the Namib Desert in terms of diet, use of space and patterns of activity to reduce competition. Co-existence is therefore possible, even in this area of low and unpredictable rainfall and food resources. Most of the food-niche separation reflects different diets. In turn, these diets are related to the exploitation of different micro-habitats (Bothma et al. 1984).

CHAPTER 5

Family Herpestidae

The smaller terrestrial carnivores have not been studied as much as the larger ones. Of the three major families occurring in the arid areas in Africa, only the Herpestidae have been studied in any depth. The herpestids include mongooses and related small carnivores. They usually live in groups, in some of which there is considerable structure and organization (Rasa 1987a; Rood 1990).

5.1 Taxonomy, Distribution, and Status

Although Skinner and Smithers (1990) still regard the herpestids as a subfamily of the Viverridae, they are considered to be an independent family by Wozencraft (1993). The Herpestidae is divided into two subfamilies, the Galidiinae, with four genera and five species endemic to Madagascar, and the Herpestinae. The entire family consists of 19 genera and 36 species. The subfamily Herpestinae is characteristic of Africa, and 21 of the 23 known species are found there. The large grey mongoose *Herpestes ichneumon* extends into Portugal, Spain, Israel, Jordan and Lebanon, and the white-tailed mongoose into the Arabian Peninsula (Fig. 5.1). No species of herpestids are listed either in Appendix I or II of CITES, but *Herpestes edwardsii, Herpestes javanicus, Herpestes smithii, Herpestes urva* and *Herpestes vitticolis* all appear in Appendix III for India. However, *Liberiictis kuhni* is considered in Liberia and the Ivory Coast to be endangered by the IUCN, while *Mungotictis decemlineata*, which is endemic to Madagascar, is vulnerable (Wozencraft 1993).

5.2 Social Behaviour

The social behaviour of the Herpestidae varies. Some species occur as solitary individuals, others live in highly integrated groups (Cavallini and Nel 1995). Sociality in mongooses may have evolved as a means of protecting the young rather than as mutual protection for the adults (Rasa 1994). Alternatively,

Fig. 5.1 The Uruq Bani Ma'arid Protection Area in the central Arabian Peninsula

habitat selection, through its effects on anti-predator behaviour and foraging strategies, may have been the leading factor in it (Cavallini and Nel 1995).

5.2.1 Group Formation

Group formation in mongooses is not correlated closely with the degree of exposure when foraging: its main selective advantage is definitely not reduction of predation pressure on the adults (Rasa 1994).

Dwarf Mongooses. The dwarf mongoose *Helogale parvula* lives in family groups throughout the semi-desert and savanna regions of Africa. As the smallest of the African carnivores, with an adult weighing some 400 g, the group is led by a dominant pair joined in a permanent bond. The rest of the group consists predominantly of their young. Within the various age groups, the division of labour and altruism is higher than in any other mammal, possibly apart from naked mole-rats. Age and sex-specific roles involve guarding, baby-sitting, and active attack on predators or conspecific intruders into the territory (Rasa 1987a).

In contrast to the majority of other social carnivores, group life in the dwarf mongoose appears to have evolved more as protection against predation on both adults and young, rather than as an efficient means of foraging

(Rasa 1987a). Group size appears to be limited by ecological pressures. Only when food is abundant and predation severely restricted, as near human habitation, are groups of 20 or more adults found. Group size is also regulated by infanticide: the killing of the young of subordinate adults. If this did not occur, group size could reach a critical level at which the territory of a pack would not contain sufficient resources to support it throughout the year. Dwarf mongooses are therefore faced with a dilemma. If the size of the pack is too small, there is no reproduction success. However, if it is too large, the vital resources in the territory are depleted. To compensate for the latter, reproduction by subordinates is suppressed by infanticide and the elimination of unsuccessful groups (Rasa 1994).

Within the framework of the restraints of predator pressure and pack and territory size in a given environment, the dwarf mongoose has had to adapt its behaviour to ensure maximum fitness at the level of the individual (Rasa 1987a). In the Serengeti National Park, dwarf mongooses live in packs with a mean size of 8.9 individuals. Each pack has a dominant breeding pair, usually the oldest individuals (Rood 1990). However, in the Taru Desert, each dwarf mongoose pack is semi-nomadic, although it lives within a territory. The mean pack size there is 12.6 (range: 2 to 28) (Rasa 1986), surprisingly larger than that in more mesic regions. The semi-nomadism of the pack seems to be a response to the distribution and abundance of its insect prey (Rasa 1989a). Entire packs in the Taru Desert become extinct if the size of the pack drops below the critical level of five individuals (Rasa 1987a).

Other Herpestids. The slender mongoose *Galerella sanguinea* of the Serengeti National Park does not live in packs. It is normally solitary, but males may sometimes form loose, non-aggresseive associations, when up to four young but adult males use the same range. Such male groups are reminiscent of the coalition of males in lions and cheetahs. The duration or tenure of a slender mongoose coalition is related directly to its size. The larger the group, the longer its duration. The largest group of males ever found was a group of four which stayed together for at least seven years. As the largest litter size of the slender mongoose is two, these coalitions cannot merely be litter mates (Waser et al. 1994).

In the arid, western sandveld grasslands of South Africa, the mean group size of the yellow mongoose is 4.1 individuals (standard deviation: 1.9) In the suricate, it is 10.6 individuals (standard deviation: 3.1; maximum: 17). The more predacious yellow mongoose is solitary, with a tendency towards sociality (Lynch 1980). Up to seven individuals may gather at a den (Grobler et al. 1984). The suricate, a mixed forager, is more social (Lynch 1980). In the central grasslands of South Africa, the mean colony size of the yellow mongoose can be as high as eight individuals (Earlé 1981). The suricate lives in large groups of up to 40 individuals which forage together (Grobler et al. 1984).

5.2.2
Communication

As vigilance behaviour, and its related system of communication, as it exists in the dwarf mongoose, are such a prominent facet of its social behaviour, this topic will be discussed separately in Section 5.2.3. The present section deals with other facets of communication known to occur in the herpestids.

Dwarf mongooses frequently use acoustic communication to check the pack's position. Visual interaction also takes place (Rasa 1989a). In the Taru Desert, vertical or horizontal objects in the territory are scent-marked with cheek and anal gland secretions in a region which is upwind of the major portion of the territory for most of the year (Rasa 1987a). The scent-marking secretions of a dwarf mongoose stay viable for up to 21 days in relatively mesic areas of South Africa (Hiscocks and Perrin 1991). This period is probably shorter in arid environments, although there are no data to prove this. Both the dwarf and the Indian mongoose *Herpestes javanicus* can distinguish between the anal gland secretions of different individuals (Rasa 1973; Gorman 1976). Under environmental conditions which preclude visual contact, acoustic signals are used to maintain contact in foraging groups (Rasa 1986). The dwarf mongoose deposits insect remains around the den (Grobler et al. 1984). However, it is not certain whether this has a communicatory function.

The Cape grey mongoose is endemic to southern Africa's more arid regions, but is absent from areas where the rainfall is less than 200 mm per year (Cavallini 1992). It differs from other herpestids in that it does not make faecal piles. Its droppings are usually found singly or in small groups (Cavallini1992; Cavallini and Nel 1995). Faecal middens are characteristic of the boundaries of a territory among slender mongooses living in the grasslands of central South Africa. When scent-marking, the dominant male often moves from one midden to the next. Anal and cheek gland secretions are used in territorial marking (Earlé 1981), which is carried out by all individuals. Marking is considered to be a possible means of sexual advertisement by subordinate individuals as they mark much more frequently than do the dominant males (Wenhold and Rasa 1994). In the arid Postberg Private Nature Reserve, the faeces of the slender mongoose is deposited in large middens close to the dens (Cavallini and Nel 1995).

The yellow mongoose uses the conspicuous white tip of the tail to communicate its position to conspecific individuals, especially in bushy areas (Earlé 1981). It leaves dung piles around the burrow entrances and at several latrines away from the burrow (Grobler et al. 1984). The behavioural repertoire of captive suricates *Suricata suricatta* includes leg-lifting (anal secretion), body-rubbing, and sniffing. These could all serve as methods of communication to orient or reassure other individuals in the pack. They may also provide each pack member with a characteristic group odour that promotes recognition and familiarity within an exclusive social group. Furthermore, anal-dragging, which follows defecation in 37% of cases, may have a social communicatory and grooming function (Moran and Sorenson 1986).

Banded mongooses *Mungos mungo* use the senses of sight and hearing, which are acute, to locate each other when foraging. Scent-marking is rare when foraging, but frequently occurs when drinking. The peri-anal glands are used in scent-marking. Males scent-mark more often than females. Scent-marking is also used as displacement behaviour during courtship (Simpson 1964).

5.2.3 Vigilance Behaviour

Vigilance behaviour has been studied extensively in birds. Information is transmitted within a foraging group of birds by acoustic signals which relay specific messages. A signal of this kind is called a "watchman's song" (Rasa 1986). It is now known from studies on dwarf mongooses in the Taru Desert, that male dwarf mongooses use a similar system through which they communicate complex information, most often when the pack is foraging (Beynon and Rasa 1989).

With their attention focused on the ground, feeding or foraging animals cannot be alert for predators. Feeding and foraging must therefore be interrupted frequently to scan the environment. When the need to do this becomes excessive, foraging may be hampered. Among social animals and species which congregate while foraging, this constraint can be reduced when more than one individual participates in vigilance behaviour. Dwarf mongooses do this. Vigilance behaviour is a sex-based, social role almost exclusive to subordinate, male dwarf mongooses. Among them it has developed into a complex system which incorporates a variety of coordinated parameters, including guarding behaviour (Rasa 1989a,b).

Rasa (1987b) pointed out that the "watchman's song" of the dwarf mongoose has evolved in an area of savanna bush with tall grass undergrowth. These conditions preclude visual contact between pack members, which are exposed to group predation when foraging. Coordinated vigilance is the rule among dwarf mongooses in the Taru Desert. The ratio of guards to adult group members is 1.7:10 (Rasa 1986). Guarding may be the most important method by which predation on these dwarf mongooses is reduced. As predation is the major cause of death amongst the mongooses (Rasa 1987b), it becomes an important survival strategy.

Many aspects of guarding behaviour are innate, but others are learned by imitation. Guarding behaviour has an extremely high learning component. Not only do the identities of the major predators of dwarf mongooses need to be recognized by the guards, but they have to be learned specifically (Rasa 1987b).The effectiveness of a group depends upon its size. When pack size

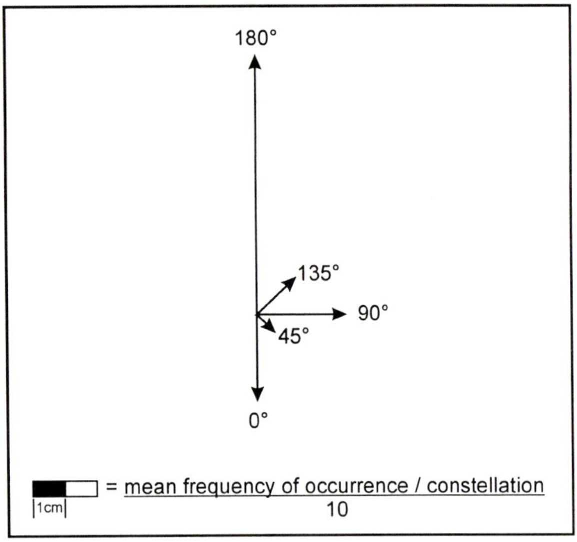

Fig. 5.2. The relationship between the direction in which a dwarf mongoose guard in the Taru Desert in Kenya looks, and the foraging direction of the pack. The various vector arms are drawn to scale to show the mean frequency of occurrence of each type of guard/group constellation divided by 10. The figure 0° indicates that the guard is looking in the direction in which the group is travelling, 180° indicates that the guard is looking directly away from the group, while 45°, 90° and 135° are transitional phases in this basic pattern (Rasa 1989a)

becomes so reduced that there are too few guards, the vigilance system is inefficient (Rasa 1989b). As pack size decreases, the subordinate males increase their total guarding time, but the time spent in guarding by the dominant and the subordinate females shows no change (Rasa 1987b).

Guard posts must be at least 1 m above groundlevel to give the guard a clear view over the grass and undergrowth. Termite mounds and trees are the guard posts most often used. Guard posts are not selected randomly, but are distributed evenly throughout the territory. They are probably used traditionally by a given pack. Visual communication as well as acoustic signals take place between the guards. The guards mainly act as rearguards, maintaining vigilance at an angle of 180° to the direction in which the group forages. However, not all guards are to be found in the rear, and some also guard the front. The latter run ahead of the group to take up position from which to look back in the direction in which the group is foraging (0° angle), at right angles (90°), or occasionally elsewhere (45° and 135°) to the direction of travel (Fig. 5.2.). The guards change regularly on a rotational basis, and the whole guarding sequence is shared equally (Rasa 1987b, 1989a).

While foraging, and especially so in dense, tall grass, dwarf mongooses maintain contact by distinctive calls. These differ in pitch between individuals. The guards also give contact-maintaining calls which increase in volume the further the group moves away from the guard emitting the call. When the pack moves too far away, another guard takes over from the first. The smaller the size of a pack, the greater is the energy expended by its guards (Rasa 1986).

When guards are present, the dwarf mongooses of the Taru Desert are exposed to predator attacks at a mean rate of 0.13 per h for aerial attacks, and

0.0016 per h for ground attacks. Ground attacks are normally more successful than aerial attacks. However, when they are detected in time, the predator invariably gives up the attack. Of all aerial attacks, 93% are unsuccessful. Vigilance is therefore highly successful. The protection afforded to packs of five animals or less is low, but it increases as the pack becomes larger, reaching an ideal in packs of between 12 and 15 members. The mean pack size of 12.6 individuals found in the Taru Desert is therefore in the optimum range for effective guarding. The formation of such large packs also reduces the work load on the subordinate males which are most involved in guarding (Rasa 1989b).

A guard that sights a predator warns the pack by any of a variety of calls, depending upon the type of predator sighted (Rasa 1986, 1989a). Only those raptors that prey on dwarf mongooses elicit warning calls, and these calls vary in type and intensity, depending upon the species of raptor concerned, its position and its danger to the pack (Rasa 1987b). The warning call also contains information on the degree of danger, the distance between the predator and the pack and, in the case of aerial predators, the elevation of the predator (Beynon and Rasa 1989; Rasa 1989a).

The vocal codes used are modified according to the type and location of the predator. They elicit an appropriate response in the pack (Beynon and Rasa 1989). The codes used are differentiated by means of frequency modulation. Distance is indicated by two pulsed calls which always precede frequencey modulated calls. Calls also vary in frequency, duration and number of pulses. In using these calls to relay specific predator-related information, the calls are compatible with the criteria of a language (Beynon and Rasa 1989).

5.3 Reproduction

5.3.1 Dwarf Mongooses

Only the socially dominant male and female in a pack of dwarf mongooses raise young, although subordinate females may mate with the dominant and subordinate males (Rasa 1987a; Creel et al. 1992). Reproduction by subordinate pack members is suppressed, and this prevents the pack size from getting out of balance with the available resources (Rasa 1994). Mating among subordinate females is suppressed by a complex of endocrinal and behavioural mechanisms. Mating in subordinate males is mainly suppressed by behavioural mechanisms (Creel et al. 1992). By means of careful division of labour and resources, dwarf mongoose packs of eight or more members in the Serengeti National Park raise twice as many young to the yearling stage than do packs of less than eight (Rood 1990).

Births occur mainly from November to April, when most of the rains fall, and when insects, the principle food resource of dwarf mongooses, are most abundant (Rood 1990). Mating lasts from 1 to 7 days, a period during which the dominant male guards the dominant female intensively. There is a significant ranking correlation between mating individuals, and age has a greater impact on sexual dominance than body mass. The dominant female usually dominates the dominant male. The presence of a dominant pair is relatively unimportant in determining whether subordinate females will mate. Nevertheless, aggressive interference by the dominant pair will often prevent subordinates from mating (Creel et al. 1992). Some subordinate males and females obtain breeding status by emigrating and forming new packs, while a subordinate male occasionally displaces the dominant male of the pack (Rood 1990).

Gestation lasts from 56 to 58 days (Rasa et al. 1992), and litters of up to 4 are born (Grobler et al. 1984). No nest is built for the young because the young are moved regularly from one termite mound to another (Rasa 1987a). The ventilation shafts of these mounds are used as dens by resting dwarf mongooses (Rasa 1989a). The young are moved at a mean interval of once every 1.6 days during the first 3 weeks of life (Rasa 1987a). A young dwarf mongoose will remain in a den for the first 6 weeks of its life, where it is guarded by one or more pack members which rotate the duty (Rood 1990). In a more mesic area of the KwaZulu/Natal Province in South Africa, however, young dwarf mongooses may stay in one den for as long as 24 days at a time.The mean duration of den tenure there is 6 days (Hiscocks and Perrin 1991). The difference with the Taru Desert may reflect resource abundance differences.

The dwarf mongoose uses helpers to share the load of raising the young, as do some canids (Jennions and Macdonald 1994). These helpers bring food to the young, or carry them from one den to another (Rood 1990). However, food is not brought to, or regurgitated for the young. Allo-parenting is usually carried out by a group of up to three helpers, depending upon the overall pack size. It usually is a female and two males. When a group of mongooses returns at midday from foraging, the helpers leave to forage for themselves, while a second set of helpers takes over the duty of looking after the young (Rasa 1987a).

The mother and the subordinate female helpers carry the young by mouth to new dens. When 3 to 4 weeks old, the young begin to accompany the group when it is foraging. Each young animal attaches itself to an adult which captures food for it, as in the case of suricates. Dominant males and high-ranking subordinate females supply food for the young. By the time they are about 4 months old, the young are able to feed themselves (Rasa 1987a; Rasa et al. 1992).

Allo-parenting has an obvious adaptive significance. There is no behavioural mechanism to provide food for a female if she stays with her young. Because the food items of the dwarf mongoose are small and difficult to find, an alternative strategy that ensures the intake of sufficient food to support lactation has evolved. Depletion of food resources around the den is also prevented by the use of temporary dens and helpers. Unguarded young, moreover, would be highly susceptible to predation (Rasa 1987a).

5.3.2 Other Herpestids

Mating in the slender mongoose resembles that of the cheetah. A female in oestrus is followed by several males of the pack. There is little aggression between them. The maximum litter size is 2 (Waser et al. 1994). The breeding season of the Cape grey mongoose coincides with the end of the rainy season (Earlé 1981; Cavallini 1992). The main prey of small to medium rodents is most abundant then. Litters of one to four young are born in a hollowed out, termite mound after a gestation period of 45 days (Grobler et al. 1984). The young are reared in rock crevices, and not in underground dens as in the case of other herpestids (Smithers 1971; Cavallini 1992).

Gestation in the yellow mongoose lasts from 56 to 58, days and is independent of body mass (Rasa et al. 1992). Litters of one to five are born in summer (Grobler et al. 1984). Two litters are produced each year within a period of 2 to 4 months in some of the arid parts of southern Africa. The first litter is born in late spring (October). In the arid Karoo region of South Africa, however, there is no second litter when food is scarce. However, in the southern Kalahari and Etosha National Park, a second period of oestrus occurs 5 to 8 weeks after the first births of the year. At this time, the young are still present and lactation is continuing (Rasa et al. 1992). The mean litter size is 1.8, and there is some evidence of induced ovulation (Lynch 1980).

The young of a yellow mongoose accompany the adults to forage when they are eight weeks old, and they remain dependent upon them nutritionally until they are 16 to 18 weeks old (Rasa et al. 1992). The yellow mongoose brings larger food items such as rodents, bats, reptiles and large arachnids to the young while they are still in the burrow. This behaviour is typical of social canids too (Rasa et al. 1992).

The yellow mongoose has apparently evolved a unique breeding strategy in which small and relatively sedentary litters follow each other rapidly in time. This helps to cope with constaints such as variable food resources and predation pressure in an arid and fluctuating environment. In the Karoo region, where no second litter is produced when food is scarce, a high level of nutritional investment in the young is spaced over a longer period. Predation pressure is also reduced, resulting in enhanced survival rates of young (Rasa et al. 1992).

The reproduction period in the suricate is long, but is not synchronized in the same population. Consequently, females in all stages of the reproductive cycle are found within a population throughout the year (Lynch 1980). However, in a particular pack, only one female breeds, while the others help to feed the young (Mills et al. 1984). The mean litter size is 2.9 in the western grasslands of central South Africa, and there is some evidence of induced ovulation (Lynch 1980). Up to five young are born per litter, after a gestation period of 60 to 70 days. The young are born with closed eyes and the mother brings food to her young at the den (Grobler et al. 1984).

The banded mongoose has a gestation period of some 60 days, as does the Egyptian mongoose *Herpestes ichneumon*. In the banded mongoose, the females are sexually mature at nine to 10 months of age. Litters of up to eight young are born after a gestation period of 60 days (Grobler et al. 1984). The eyes are closed at birth, a time when the young are covered with fine hairs. The eyes open when the young are 10 days old. The adult pelage appears when they are 45 days old. The young are not fed solid food in the den. They begin to accompany the foraging adults at an age of 3.5 to 4 weeks (Rasa et al. 1992). Single young animals accompany single adults who capture food for them (Rasa 1987a).

5.4 Area Use, Movements, and Activity

5.4.1 Dwarf Mongooses

The territory of a pack of dwarf mongooses in the Taru Desert may contain up to 200 or more termite mounds (Rasa 1987a). The ventilation shafts of these are used extensively as dens by the mongooses. The size of a territory varies between 0.65 and 0.96 km² (Rasa 1989a). There is little overlap between territories, and no correlation between pack and territory size. Dwarf mongooses do not use larger territories than necessary. They maintain the smallest economically defensible area which contains sufficient food. These territories are stable over many years (Rasa 1987a). The dwarf mongoose of the Taru Desert is active by day, sheltering in the ventilation shafts of termite mounds at night and during the heat of the day (Rasa 1989a).

Most of the herpestids shelter in dens of one kind or another. The dwarf mongoose of the Taru Desert shelters in the ventilation shafts of mounds of *Macrotermes* and *Odontotermes* spp. at night and during the heat of the day, as well as to hide from attacking predators. These mounds are abundant, and spaced relatively evenly at a density of about 1 per 25 m² (Rasa 1989a). This means that a given dwarf mongoose territory will contain 200 or more termite mounds (Rasa 1987a). This ensures that a potential refuge from predators is always close at hand. As a new termite mound is selected for sleeping in every night, predation pressure is spread evenly over the territory of a given pack of dwarf mongooses (Rasa 1987a).

The preferred type of termite mound for denning in KwaZulu/Natal seems to be a multi-holed termitarium located in a thicket. Dens of the dwarf mongoose are often to be found in *Odontotermes badius* mounds. Each particular den is used for one night only, although it will be re-used later. After repeated use, a den may become unsuitable and will then not be re-used later. Nevertheless, 56.9% of all dens are re-used by dwarf mongooses. The availability of suitable termite mounds for dens forms a vital part in habitat selection by both the dwarf and banded mongooses which use the same dens, but not simultaneously. The environmental features around a mound are important in determining its suitability as a den. The nature of the vegetation determines the degree of protection afforded against aerial attack. The ideal den for the dwarf mongoose is a multi-entranced termite mound covered with an open thicket at a mean distance of 4 m away from the nearest dense cover (Hiscocks and Perrin 1991).

5.4.2 Slender Mongooses

The range of the slender mongoose in the Serengeti National Park may be used by up to four males. These form loose, non-aggressive associations which may last for many years. The relationship between the occupants of a range is amicable (Waser et al. 1994). The slender mongooses of the more mesic grasslands of South Africa do not rely exclusively on natural holes, hollow logs and the burrows of other animals for dens. They are adept diggers although they do not have strong claws, and are capable of excavating their own burrows if the need arises (Bronner 1992). They stay in their dens on cold days, and when they move about they tend to use tracks and footpaths (Smithers 1971).

5.4.3 Cape Grey Mongooses

Cape grey mongooses in the arid regions of southern Africa occupy ranges which overlap greatly, especially among males. These ranges vary in size in summer between 21 and 63 ha (Cavallini 1992; Cavallini and Nel 1995). The range sizes are many times greater than that of the dwarf mongoose in the Taru Desert (Rasa 1989a), or even the 5 to 6 ha range of the yellow mongoose in central South Africa (Earlé 1981). As the range of the yellow mongoose is patrolled along a definite route, it appears to be a true territory. Faecal middens are used to scent-mark the boundaries (Earlé 1981). In the Postberg Private Nature Reserve, an area with an arid, mediterranean climate, the mean range size of a yellow mongoose is 84 ha (standard deviation: 31 ha) (Cavallini and Nel 1995).

The Cape grey mongoose does not shelter in a burrow in summer. It uses a rock crevice where the young are also reared (Cavallini 1992). When a burrow is used in winter, it is often in an old anthill, but occasionally under the roots of large trees, in hollow logs or in broken, rocky areas (Simpson 1964).

5.4.4 Yellow Mongooses and Suricates

Both the yellow mongoose and the suricate are diurnal. Especially in summer, they rest underground in their dens during the period of maximum temperature outside. During the winter, when day temperatures are low, they emerge from their burrows to sunbathe.

In the arid, western grassland areas of central South Africa the yellow mongoose and the suricate spend 16 out of each 24 h underground during summer, and 12 of 24 h underground in winter. Although day-active, the yellow mongoose also forages at twilight and briefly at night. In contrast, the suricate is strictly diurnal in its foraging habits (Lynch 1980). Onset of activity in the yellow mongoose is regulated by the ambient temperature (Earlé 1981). The yellow and Cape grey mongooses of the arid, mediterranean Postberg Private Nature Reserve are both diurnal, but only the yellow mongoose uses an underground den (Cavallini and Nel 1995): the Cape grey mongoose shelters in rock crevices (Cavallini 1992).

The yellow mongoose is social, and males and females den together (Cavallini and Nel 1995). In the western sandveld regions of South Africa, the yellow mongoose excavates its own burrows. It also frequently inhabits the burrows of the suricate and the ground squirrel *Xerus inaurus* sympatrically. Both use simple burrows, with a few entrance holes, and more complex burrows, with multiple entrances. The burrow system is dynamic and is constantly cleared of debris. In the process it is enlarged. Different burrow systems are 50 to 100 m apart and up to five such systems may exist within the range of a given yellow mongoose family or suricate pack, although only one system is occupied at a given time. The systems available in a given range are used in rotation. The interval between successive periods of use depends upon the availability of food, the density of the mongooses, environmental conditions (especially flooding), and the presence of predators.The habitat type does not seem to influence the location of a burrow system, but there is a preference for calcareous soils. The usual burrow system has a mean diameter of 5 m with 15 entrance holes, but as many as 66 entrance holes are known for the yellow mongoose. An entrance hole has a mean diameter of 150 mm, and angles down at 40° from the horizontal. Tunnels are usually 75 mm high, and slightly oval. They run 1 to 1.5 m underground in two to three levels (Lynch 1980).

The burrows of both the yellow mongoose and the suricate are co-inhabited by various other small mammals, including murids, the rock hyrax, the slender mongoose, the Cape grey mongoose, reptiles, amphibians and even birds. The burrow is well-insulated against thermal extremes. Daily air temperatures on the ground surface in the arid, western sandveld areas of South Africa fluctuate by 18 °C in summer, and 30 °C in winter, whereas the mean daily burrow temperature fluctuates only by 0.7 °C in summer, and 0.8 °C in winter. The mean burrow temperature in winter is 13.3 °C colder than it is in summer (Lynch 1980).

As they often occupy burrows sympatrically, the burrow of a suricate is similar to that of a yellow mongoose. A suricate does not always excavate its own burrow, but may use one of another species, such as the ground squirrel. A single suricate pack may occupy two adjacent burrow systems. Some packs use several burrow systems in a short period of time, while others use the same system for several years. A suricate burrow system has a mean of 15 entrance holes, but can have as many as 90 (Lynch 1980). A suricate does not leave its burrow until well after sunrise, returning well before sunset. It also does not leave the burrow on cold, overcast or rainy days (Smithers 1971).

5.4.5 Banded Mongooses

In the more mesic KwaZulu/Natal Province, banded mongooses also use termitaria as burrows. Multi-entranced termitaria close to cover are used, and preferably in open thickets. However, the ideal burrow of the banded mongoose is less dependent upon vegetation physiognomy than that of the dwarf mongoose. Of the burrows used, 35% are used more than once (Hiscocks and Perrin 1991).

5.5 Feeding Ecology

The evolution of group life in carnivores has been attributed to a variety of ecological conditions. Most of them are associated with the capture and safeguarding of prey large enough to feed several individuals. However, this does not explain the development of group life among smaller species. Here, protection of feeding group territory seems to be the main evolutionary driving force. The tendency among social mongooses suggests that renewability of food resources is more important than the size of the prey (Rasa 1994).

5.5.1 Food

Insects are the main prey of the dwarf mongoose (Rasa 1989a; Rood 1990). The prey consists mainly of small, arthropods, but also includes an occasional small mammal, lizard or small bird. Such food is scattered and unpredictable, and to obtain it requires extensive searching (Rasa 1987a).

The slender mongoose eats insects, scorpions, reptiles, small birds, murids and even wild fruits (Smithers 1971). The suricate eats a wide variety of prey, much like the dwarf mongoose (Lynch 1980; Rasa 1987a). In addition, it also eats plant material, mostly seeds (n = 21 stomachs) (Zumpt 1968). Based upon the contents of 98 stomachs examined from the western grasslands of central South Africa, the suricates there eat Coleoptera (58% of all stomachs), Lepidoptera (43%), Isoptera (40%: all *Trinervitermes trinervoides*), Orthoptera (34%), Diptera (23%), Arachnida (21%), Hymenoptera (15%), Dermaptera (12%), Chilopoda (10%), Dictyoptera (10%), Diplopoda (9%), reptiles (5% : mainly skinks), amphibians (5%), birds (2%) and Hemiptera (1%) (Lynch 1980).

The yellow mongoose is also predominantly insectivorous, but more carnivorous than the suricate (Lynch 1980). Its diet includes rodents, birds, bird's eggs, lizards, and even wild fruits and carrion (Zumpt 1968; Smithers 1971; Herzig-Straschil 1977; Grobler et al. 1984). In the western grasslands of central South Africa, the termite *Hodotermes mossambicus* is an important food for the yellow mongoose, and some stomachs contain nothing but these termites. The mammals eaten are mainly murids, while a few ground-nesting birds are also taken. Of 63 stomachs examined, 67% contained invertebrates and 28% contained vertebrates (Herzig-Straschil 1977). In winter in certain areas, wild fruits are used opportunistically as food too (Earlé 1981). In the southern Kalahari, the yellow mongoose also eats the venomous *Parabuthus* sp. of scorpion (Viljoen and Davis 1973). Of 156 stomachs of the yellow mongoose examined from the western, sandveld areas of South Africa, 74% contained Isoptera (all *Trinervitermes trinervoides*), followed by Orthoptera (48%), Coleoptera (42%), Lepidoptera (22%), birds (11%), Arachnida (11%), amphibians (10%), mammals (9%: mainly murids, cricetids and the springhare *Pedetes capensis*), Hymenoptera (9%), reptiles (6%), Dermaptera (6%) and Dictyoptera (3%) (Lynch 1980).

Even in the arid, southern mediterranean fynbos (macchia) vegetation of the Postberg Private Nature Reserve, where rodents are abundant, the yellow mongoose maintains a mainly (65% occurrence) insectivorous diet, as opposed to the Cape grey mongoose whose diet there consists of 95% rodents (Avenant 1993; Cavallini and Nel 1995). Insects are eaten secondarily, and the Cape grey and yellow mongoose also scavenge for carrion (Cavallini 1992). Nevertheless, the Cape grey mongoose is a generalist and opportunistic feeder. Young birds and bird's eggs are eaten mainly in spring, the breeding season of the birds. Adult birds are taken throughout the year. Reptiles are eaten especially during the summer (Avenant 1993). Even the venomous boomslang *Dispholidus typicus* is killed and eaten (Cavallini 1992).

The slender mongoose is omnivorous, but mainly eats insects. It also eats carrion, small rodents, bird's egss, other invertebrates and occasionally lizards and snakes, including the venomous cobra *Naja* spp. and young mambas *Dendroaspis polylepis*. The banded mongoose feeds mostly on invertebrates, but they also eat reptiles, small mammals and birds. Where they occur in coastal dunes, they also eat snails. The shell of a snail is broken as is done with eggs by other mongooses: the mongoose flicks the egg with the front paws through the hind legs against a solid object. More than 80% of the diet of the suricate consists of insects, mainly beetles. However, suricates also feed on reptiles, frogs, spiders, scorpions, small mammals, birds and even wild fruits (Grobler et al. 1984).

In the northern Mediterranean, the small Indian mongoose *Herpestes javanicus* of Korcula Island in the Adriatic Sea, eats an inordinately large proportion (43% by volume) of plant material, mainly fruits. Its other food consists 46% by volume of vertebrates, mainly murids. The rest of its food consists of invertebrates and various other items. However, with the exception of shrews, the entire small mammal fauna of the island is eaten. This mongoose is a generalist there, and not insectivorous as has been previously believed. The fruits eaten are mainly those of the juniper *Juniper oxycedrus* and the strawberry tree *Arbatus unedo*. Of vertebrates, wood mice *Apodemus* spp. and rats *Rattus domesticus* predominate in the diet. Passerine birds, reptiles and invertebrates (7.5% by volume of Orthoptera and Coleoptera) are also eaten. The large amount of fruits eaten is advantageous in that it provides additional sugar and fats which aid these mongooses in heat production during the cold, wet winter (Cavallini and Serafini 1995).

5.5.2 Hunting and Foraging

The semi-nomadic dwarf mongooses of the Taru Desert forage extensively over relatively long distances every day, but each individual in the pack hunts and kills for itself (Rasa 1987a). Dwarf mongooses forage by searching through leaf litter and stony ground. Eggs are broken as in other mongooses (see above; Grobler et al. 1984). Large packs move over significantly longer distances than smaller ones, a phenomenon which is related directly to energy input (Rasa 1987a). While foraging, contact is maintained between individuals of the pack by individually identifiable contact calls, as stated earlier (Sect. 5.2.3; Rasa 1986).

In contrast, the yellow mongoose of the arid, western sandveld areas of southern Africa hunts alone. It scratches around a lot, digging up beetle larvae in the process. Individual yellow mongooses vary in foraging approach and some tend to scratch in the vicinity of zebra or steenbok dung piles. However, the biggest single source of food is termites (Earlé 1981). When foraging or hunting, the yellow mongoose is quiet in contrast to the highly vocal suricate which often occurs sympatrically with it (Lynch 1980). The yellow mongoose is phylogenetically closer to ancestral carnivores than is the suricate. It appears to have attributes intermediate between those of the extinct miacids and recent social mongooses (Lynch 1980). The yellow mongoose has retained its ability to catch small rodents, presumably the normal prey of ancestral mongooses, but it is more insectivorous. Its main diet, however, is probably an adaptation to the higher availability of insects in the open fields which it now inhabits. Dietary, behavioural and morphological differences may therefore be secondary here to differences in habitat selection (Cavallini and Nel 1995). The banded mongoose forages by turning over stones, and digging in leaf litter and other debris. The claws of the front feet are used when digging and scratching around (Grobler et al. 1984).

The suricate of the arid, western sandveld parts of southern Africa scratches and digs more when foraging than does the yellow mongoose. Its long front claws are adapted for digging (Smithers 1971). Scent is important for locating food underground. When foraging, the suricate digs, rakes, or moves litter around. Fresh animal droppings are examined for insects, a major part of the diet (Simpson 1964). Suricates do not forage further than 500 m away from their burrows (Smithers 1971).

5.5.3. Killing, Feeding, and Water Use

No cooperative killing of prey occurs in dwarf mongooses, because they are not able to kill large prey, and there is no cooperative hunting. Food sharing also does not occur. Each individual hunts and kills for itself (Rasa 1987a). In contrast, both the yellow mongoose and the suricate have dentition well-adapted to kill prey as large as small rodents (Cavallini and Nel 1995). However, the long canines may have a social function too (Grobler et al. 1984). The banded mongoose uses its claws to hold larger prey onto the ground, but smaller food items are eaten whole. Little data are available on the water-dependency of mongooses. Considering their distribution and habitat choice (Dorst and Dandelot 1970; Smithers 1983), however, many species are probably independent of free water. Furthermore, it is known that the banded mongoose will drink water early in the morning or at dusk (Simpson 1964). The yellow mongoose probably eats succulent plants for moisture (Zumpt 1968).

5.5.4 Population Ecology

The population ecology of most herpestids is poorly known. The survival rate of juvenile and adult dwarf mongoose of the Serengeti National Park increases with pack size (Rood 1990). In the Taru Desert, this phenomenon is probably related to vigilance behaviour and guarding (Rasa 1987b). Small dwarf mongoose packs in the Taru Desert are unable to raise young on account of predation pressure. The dwarf mongooses there, however, reduce predation risk by sleeping in a new termite mound den every night, spreading predation pressure over the entire range in the process (Rasa 1987a).

The sex ratio of a yellow mongoose population in the arid, western sandveld of South Africa is 1.2 females per male (Lynch 1980). A young yellow mongoose accompanies the adults to forage from the age of 8 weeks when it is able to avoid raptors. Young yellow mongooses are mainly killed by terrestrial predators, especially snakes and monitor lizards (Rasa et al. 1992). The mean sex ratio of the suricate is 1.09 males per female (Lynch 1980). In response to every killing campaign of yellow mongooses and suricates for rabies control in South Africa, there has been an enormous increase in insects and rodents in these areas. Eventually, the numbers of birds increased, which in turn limited the insect populations. Emigrating yellow mongooses and suricates soon moved into the areas that became vacant due to such control actions, populated them, and started a new population cycle of 2 to 3 years' duration (Zumpt 1968).

5.5.5 Interaction with Other Carnivores

Little is known about the interaction between herpestids and other carnivores. The yellow mongoose and the suricate often inhabit the same burrow, but they do not compete for it. Although they are both insectivorous, they select different categories of insects as food. Furthermore, they feed at different times. The yellow mongoose is more crepuscular, while the suricate is day-active (Lynch 1980). The suricate is preyed upon by large raptors (Smithers 1971). Interspecific competition with other small carnivores is limited by niche separation. It is most severe with the closely-related slender mongoose (Cavallini 1992).

In the southern Kalahari, the greatest overlap in common prey in the diet is between the yellow and the slender mongoose, and between the slender mongoose and the striped polecat. However, the yellow mongoose and the striped polecat show little dietary overlap. Furthermore, the yellow and slender mongooses are diurnal and the striped polecat nocturnal, reducing the degree of exploitative competition. There are also differences in habitat selection which reduce competition (Mills et al. 1984).

CHAPTER 6

Families Mustelidae and Viverridae

Fewer data exist on the mustelids and viverrids of arid areas than on the herpestids. Only a brief and selected survey will therefore be given here, to serve as an example of what can be expected.

6.1 Family Mustelidae

6.1.1 Taxonomy, Distribution, and Status

The mustelids are distributed widely throughout the world, but they are not found in Australia, Madagascar or the polar regions. Those that occur partially or entirely in the arid areas of Africa include the honey-badger or ratel *Mellivora capensis* whose range extends into Asia, the common or least weasel *Mustela nivalis* of Egypt, Algeria and Morocco, the stoat or ermine *Mellivora capensis* which extends into the extreme north of Africa, the ferret *Mustela putorius* of northern Morocco, the snake mongoose or white-naped weasel *Poecilogale albinucha* of Zaire and Uganda south to South Africa, the North African striped weasel *Ictonyx libyca* from the fringes of the Sahara Desert in Morocco and Egypt to Mauritania, northern Nigeria and Sudan in the south, the zorilla or striped polecat *Ictonyx striatus* from Senegal, south-eastern Egypt and Ethiopia, and south to South Africa (Dorst and Dandelot 1970; Wozencraft 1993).

The Mustilidae currently consists of the five subfamilies Lutrinae, Melinae, Mellivorinae, Mephitinae and Taxidiinae. Of the 25 genera and 63 species recognized, only 6 are restricted to Africa. They are the Cape clawless otter *Aonyx capensis*, the Zaire clawless otter *Aonyx congicus*, the spotted-necked otter *Lutra maculicollis*, the North African striped weasel *Ictonyx libyca*, the zorilla or striped polecat *Ictonyx striatus* and the snake mongoose or white-naped weasel *Poecilogale albinucha*. Only the honey-badger occurs both in Africa and elsewhere (Wozencraft 1993).

Of the 63 known species, 15 are on either Appendix I or II of CITES. Of the Lutrinae are: *Amblonyx cinereus, Aonyx capensis, Aonyx congicus, Enhydra lutris, Lontra canadensis, Lontra felina, Lontra longicaudis, Lontra provocax, Lutra lutra, Lutra maculicollis, Lutra sumatrana, Lutra perspicillata* and *Pteronura brasiliensis*. None of the Melinae appear on this list, but *Mellivora capensis*, the only species of the subfamily Mellivorinae is listed in Appendix III for Ghana and Botswana. Of the Mephitinae, *Conepatus humboldtii*, and of the Mustelinae, *Mustela nigripes* are listed in Appendix I or II. In addition, Appendix III contains *Eira barbara* (Honduras), *Galictis vittata* (Costa Rica), *Martes flavigula* (India), *Martes foina* (India), *Martes gwatkinsii* (India), *Mustela altaica* (India), *Mustela erminea* (India), *Mustela kathiah* (India) and *Mustela sibirica* (India). The IUCN considers the species *Lontra felina, Lontra provocax, Pteronura brasiliensis, Gulo gulo* and *Mustela lutreola* and the subspecies *Lontra longicaudis longicaudis* and *Lutra lutra lutra* as vulnerable, while *Mustela nigripes* of North America is considered to be endangered. In the Amazon Basin of Brazil, Ecuador and Peru *Mustela africana* is of interest because its original type locality is erroneously listed as Africa, hence its name. However, it does not occur in that continent (Wozencraft 1993).

6.1.2 The Honey-Badger

The honey-badger, or ratel as it is also called, will be discussed briefly as an example of the Mustelidae in an arid region. It occurs extensively in such areas, although it is also found in mesic forests and grasslands (Walker 1983; Skinner and Smithers 1990). The generic name *Mellivora* is derived from the Latin *mel* (honey) and *voro* (to devour), and aptly describes a dietary predilection of this carnivore (Skinner and Smithers 1990).

Honey-badgers are solitary or move around in pairs (Smithers 1971; Goss 1986). There are no solid data on reproduction, although the gestation period of 6 months and a litter size of two to three are given by Walker (1983) and Grobler et al. (1984). The young are born underground (Grobler et al. 1984). Honey-badgers are omnivorous and their diet includes wild fruits, honey and the larvae of bees. They often eat invertebrates, of which scorpions form a high proportion. This reflects their occupancy of arid areas. Their diet also includes carnivores such as aardwolves and bat-eared foxes, murids, reptiles and birds (Kruuk and Mills 1983; Skinner and Smithers 1990). In the southern Kalahari, there is a striking dichotomy in the size of food items of honey-badgers, which corresponds with different foraging strategies (Kruuk and Mills 1983). In Botswana the large baboon spider, which hunts at night, is often dug out of its silk-lined hole in the ground and eaten (Smithers 1971). As honey and bee larvae only form 14% of their food by occurrence (Skinner and Smithers 1990) the name honey-badger is somewhat of a misnomer, but honey may be more important to it seasonally.

When hunting, the honey-badger moves with a slow, rolling gait and its nose close to the ground, sniffing for prey (Skinner and Smithers 1990). It is a prolific digger and frequently digs out rodents (Smithers 1971). In the southern Kalahari, the pale chanting goshawk *Melierax canorus* often follows a hunting honey-badger to catch any insects, rodents and other animals which may be disturbed by it (Mills et al. 1984; Skinner and Smithers 1990; Marlow 1993). Moreover, a black-backed jackal often also follows in attendance for the same purpose (Grobler et al. 1984; Mills et al.1984). This is not known in other areas. There also is good evidence of an association with the honey-guide *Indicator indicator* which helps a honey-badger to locate bee-hives and in return feeds on the destroyed hive (Smithers 1971).

Honey-badgers are nocturnal, but may forage in daytime in areas where they are not disturbed (Smithers 1971). They lie up by day in burrows such as those of aardvarks (Grobler et al. 1984). The honey-badger has no known enemies other than man (Smithers 1983; Walker 1983), although it may be occasionally killed by African wild dogs (Grobler et al. 1984). Competition for food exists in the southern Kalahari between the honey-badger and the small-spotted genet as they both feed intensively on the same three murid species (Mills et al.1984).

6.2.
Family Viverridae

6.2.1
Taxonomy, Distribution, and Status

The Viverridae consists of six subfamilies, the monotypic Cryptoproctinae endemic to Madagascar, the Euplerinae with two species endemic to Madagascar, the Hemigalinae, the monotypic Nandiniinae, the Paradoxurinae and the Viverrinae. Of these subfamilies, only the Nandiniinae and Viverrinae occur in Africa, with 12 endemic species *Nandinia binotata, Civettictis civetta, Genetta abyssinica, Genetta angolensis, Genetta johnstoni, Genetta maculata, Genetta servalina, Genetta thierryi, Genetta tigrina, Genetta victoriae, Osbornictis piscivora* and *Poina richardsonii,* while *Genetta genetta* is also found in Europe and the Middle East (Wozencraft 1993).

6.2.2
The Small-Spotted Genet

The small-spotted genet *Genetta genetta* occurs in arid, open areas for the greater part of its range. Its habits are described briefly here as an example of how members of this family behave in arid regions. It is mostly solitary or found in pairs (Skinner and Smithers 1990). Scent glands are used for communication and as a method of marking its territory. The anal glands in particular secrete a foul-smelling substance when the genet is threatened. In southern Africa, the young are believed to be born during the warm, wet season (Smithers 1971; Skinner and Smithers 1990). Their eyes are closed at birth (Grobler et al. 1984). Gestation lasts from 70 to 77 days and the mean litter size is 2.6 (range: 2 to 4) (Grobler et al. 1984; Skinner and Smithers 1990).

The young are weaned when they are 9 weeks old (Smithers 1971; Skinner and Smithers 1990). The small-spotted genet is strictly nocturnal and mainly terrestrial, although it climbs trees with ease. These genets rest by day in hollow logs, holes in trees, or in burrows such as those of aardvarks and springhares (Smithers 1971). They move slowly and deliberately when stalking, then rush in for the kill. The prey is grabbed with the sharp claws of the forelimbs and bitten repeatedly until it is dead. The five major types of food known to be eaten in southern Africa are insects, arachnids, murids, birds and reptiles in that order of priority, although other items such as frogs, centipedes and wild fruit may also occasionally be eaten (Smithers 1971; Viljoen and Davis 1973; Skinner and Smithers 1990).

CHAPTER 7

Synthesis

7.1 The Order Carnivora

Some 64 million years ago, at the end of the Cretaceous period when the dinosaurs became extinct, mammals were small and only rarely reached the size of a domestic cat. The absence of large, terrestrial carnivores after the demise of the dinosaurs created ideal conditions for the evolution of mammalian carnivores. The earliest known Carnivora first appeared in the Paleocene. They were small, arboreal viverrids. The reasons for the immense increase in diversity of the carnivores are complex, but two factors were largely involved. The first was the occupation by carnivores of Africa in the early Miocene, and of South America in the late Miocene. Intercontinental faunal exchange then led to great diversification. The second has been the occurrence of cycles of extinction and adaptation of early carnivores. For example, dirk-toothed cats evolved on at least four separate occasions, and large, bone-crushing carnivores at least five times (Martin 1989).

The earliest canid known was a small fox-like animal from the early Oligocene of North America. Modern canids started to radiate in the late Miocene in Asia. Of the felids, scimitar-toothed cats occurred in Europe, Asia, Africa and North America. The genus *Felis* apparently developed in the Miocene in North America and *Panthera* in Eurasia. All the saber-toothed cats became extinct at the end of the Pleistocene, but the conical-toothed cats only became reduced in their range. Although the hyaenids are more closely related to the felids than to the canids, they became dog-like soon after their origin in the Miocene. The herpestids first appeared in the upper Oligocene of France. The mustelids originated in the early Miocene in Eurasia and North America, but are now almost worldwide. The earliest known viverrids date from the Oligocene of Europe, but they too are now almost worldwide. However, the viverrids apparently have a long, but undiscovered history in Africa and Asia. Almost all the recent Carnivora are the result of radiations during the last seven million years (Martin 1989).

The members of the order Carnivora alive today are represented by 11 families, 23 subfamilies, 132 genera and 268 species. They are found in terrestrial, fresh water and marine habitats. Of the terrestrial carnivores, the mustelids are specifically most numerous and contain 63 species worldwide. The Hyaenidae is the smallest extant family with only four species (Wozencraft 1993). The canids occur widely throughout the world where various species occupy diverse ecological niches (Mech 1975). However, there are few canids in deep the forests (Kleiman and Eisenberg 1973). Canids are primarily dog-like carnivores designed for running. The felids are specialized carnivores, and as they occupy habitats as diverse as forests and deserts, they show great ecological adaptability (Kleiman and Eisenberg 1973; Grobler et al. 1984). Extant living hyaenids are only a remnant of what was once a large family. Hyaenids have existed for more than 25 million years (Werdelin and Solounias 1991). They are most closely related to the Viverridae and Felidae (Delany and Happold 1979; Gould 1981), although they are quite dog-like in physical appearance. The herpestids, mustelids and viverrids are a diverse group of smaller carnivores (Grobler et al. 1984).

7.2 Social Behaviour

Carnivores vary greatly in their social structure, with differences between species and even within populations of the same species (Schaller and Lawther 1969). Some carnivores are solitary, and others are highly gregarious. The strictly nocturnal species are not gregarious, although the lion and the spotted hyaena approach a truly gregarious life (Rautenbach and Nel 1978). The benefits of sociality include increased hunting efficiency and the abililty to pursue larger prey as the mass of the prey hunted increases with the mass of the individual predator or the combined mass of the group. Predation rates are therefore affected by group size (Belsky 1987). However, most carnivore species are solitary in that they never cooperate with a conspecific, except when mating. This means that they are alone when rearing young, foraging or interacting with other carnivores. In some cases, social behaviour even differs between the sexes of the same species. Solitary life is mainly the result of the absence of selection pressure for cooperation (Sandell 1989).

The greatest flexibility of carnivore social systems is reflected in area use and group formation. The origins of the diverse societies found in carnivores may well lie in the patterns of prey dispersion which have permitted some carnivores to remain together into adulthood. However, most carnivores are typically opportunist and many species have broad ecological niches, with highly flexible social systems (Macdonald 1983).

7.2.1 Group Formation

Carnivora do not usually live in groups, and only 10 to15% of all species aggregate at any time outside the breeding season. There is also no relationship between body mass and group size, but some evidence that in the carnivores which eat mainly meat, group size may be related to prey size (Gittleman 1989). There is no correlation between group size and territory or range size among carnivores. In those groups that do form, the group consists largely of related individuals (Macdonald 1983).

Group size does not correspond strictly to prey size in canids (Moehlman 1989). Specializations which have occurred in group living canids serve to maintain group cohesion. For example, the social ranking among African wild dogs revolves around a dominant male and female which lead the pack (Mech 1975). The more solitary species are usually more fox-like, and the social ones more dog-like - with the exception of the bat-eared fox which is somewhat social (Kleiman 1967). Pack size varies with the resources available. The Simien fox also occasionally forms small packs when hunting, as does the golden jackal (Wandry 1975). Large groups of these jackals may form at food in Israel (Müller-Using 1975; Ferguson et al. 1983; Khidas 1990). The side-striped and black-backed jackal live mainly in monogamous pairs, which may be accompanied as helpers by the young of the previous litter. However, at localized sources of abundant food, such as the seal rookeries along the Namib Desert coast, up to 60 black-backed jackals are known to congregate in a single group (Ferguson et al. 1983). Depending upon the resources available, the group size of the bat-eared fox in the southern Kalahari varies greatly over the years. There is a linear increase in this region in the frequency of occurrence of groups larger than two as rainfall, and hence food resources, increase (Nel et al. 1984).

A solitary life is compatible with felid behavioural specializations (Sunquist and Sunquist 1989). All felids but the African lion, and to some degree the cheetah, live a largely solitary life (Schaller 1972; Caro 1994). Group size in the lion may be larger when conditions are favourable, but it is usually adapted to the worst of recent conditions (Macdonald 1983). Consequently, pride size generally tends to be lower in arid than in mesic areas (e.g. Schaller 1972; Eloff 1973a; Caraco and Wolf 1975).

In some Hyaenidae such as the brown hyaena, resource availability and dispersion create conditions under which the smallest range that can support a pair of animals, will also be able to support a group of them. The size of a spotted hyaena clan varies greatly between habitats from large, stable groups to small transient ones. This influences almost every aspect of social behaviour (Macdonald 1983). In the more social herpestids, such as the dwarf mongoose, maximum pack size appears to be limited by ecological conditions (Rasa 1994), and minimum pack size by the ability to guard the pack properly from predator attacks while it is foraging (Rasa 1989b). Pack size in the dwarf mongoose has developed more as a protection against predation than for improved foraging (Rasa 1987a). Some of the herpestids do not live in packs, but the slender mongoose may form loose male associations (Waser et al. 1994).

Nomads are rare among canids (Boggs and McNutt 1994). However, in felids and hyaenids they play a vital role in reproduction (e.g. Schaller 1972; Bertram 1979; Mills 1990; Hofer and East 1995). Nomadism is apparently also rare in the herpestids, mustelids and viverrids. A roaming tactic is more profitable the more dominant the male is, since the ability to take over females is related to the male's social status (Sandell 1989).

7.2.2 Communication

Carnivores live complex, social lives in which order is maintained by visual, olfactory and auditary signals (Gorman and Trowbridge 1989). These signals vary greatly between species. Scent-marking in particular can convey complex messages, especially to conspecifics. However, it also orientates the scent-marking individual in its own environment. It includes the use of urine, faeces and glandular secretions, although there probably is not a specific pheromone associated with scent-marking (Johnson 1973). In general, males scent-mark more frequently than females, often in response to reproductive stimuli (Johnson 1973; Rieger 1979b). Scent-marking is not analogous in all carnivores (Johnson 1973), and may be restricted to territorial species (Rieger 1979b). It is, however, complex behaviour which can have many functions (Johnson 1973).

Canids generally employ a wide range of similar postures and movements to communicate (Kleiman 1967). However, they use vocalizations and scent-marking too (Kleiman 1966, 1968). Social canids rub scents on themselves more than solitary canids do (Rieger 1979b). Canids scatter faeces singly or in latrines throughout their range (Gorman and Trowbridge 1989). As most felids live solitary lives, vocal and visual communication among them has developed to varying degrees. However, complex, expressive movements are highly developed (Schaller 1972). Most felids defecate along trails and onto elevated objects, and claw trees. Males, especially, spray-urinate backwards so that anal secretions mix with the urine (Smith 1978; Bothma and Le Riche 1984; Gorman and Trowbridge 1989). Felids also scrape on the ground for chemical communication (Verberne and Leyhausen 1976). When scraping, interdigetal gland secretions are deposited.

Hyaenids use a variety of tactile, vocal and olfactory systems of communication (Gorman amd Trowbridge 1989; Mills 1990). However, urine does not appear to have any communicatory function (Owens and Owens 1978). Pasting, as defined in Section 4.2.3, and the use of faecal latrines play a major role in scent-marking (Bearder and Randall 1978; Mills 1990). The whooping call of the spotted hyaena is used to maintain intergroup spacing and territory integrity (East and Hofer 1991). Moreover, it has other functions as well (Hofer and East 1995). Aardwolves and brown hyaenas appear to have two types of paste for short and long-term purposes (Nel and Bothma 1983; Gorman and Mills 1984). Brown hyaenas in the central Kalahari can even recognize each other individually from various communication signals (Owens and Owens 1978).

Herpestids employ extensive communication systems (Rasa 1973; Gorman and Trowbridge 1989). In the dwarf mongoose, vigilance behaviour has developed to such a degree that complex and detailed messages are relayed about predators to the foraging pack by the individuals who guard them (Beynon and Rasa 1989; Rasa 1989a). Scent-marking occurs in herpestids (e.g. Simpson 1964; Wenhold and Rasa 1994), and faecal middens are characteristic of species (e.g. Cavallini 1992; Cavallini and Nel 1995).The mustelids and viverrids make use of similar communication systems (Smithers 1983; Gorman and Trowbridge 1989). Intuitively, scent signals might be thought to be less effective in an arid than a mesic environment. However, no data exist regarding this, and as the carnivores of arid regions do scent-mark extensively, it must have a role to play there.

7.3 Reproduction

Where female mammals are gregarious, and males do not defend territories, the males may either move among the females in search of receptive individuals, or remain resident with a particular group of females (Sandell 1989; Whitehead 1990). The reproductive system of carnivores varies greatly. In canids, reproduction is generally seasonal, and a copulatory tie, where the mating pair remain physically joined after ejaculation, is characteristic (Kleiman 1968; Van Heerden 1981). Obligatory monogamy occurs in most species (Moehlman 1989), even when it is linked to the dominant pair of a large pack as in the African wild dog (Bertram 1979; Frame et al. 1979; Boggs and McNutt 1994; Burrows 1995). Many canids use subordinates to help raise the young (Macdonald 1983). The bat-eared fox is also monogamous but, compared with other canids, it has small litter, and its pups are more fully developed at birth (Lamprecht 1979; Malcolm 1986). Especially among the canids, the subordinate members of a pack are reproductively suppressed (Macdonald 1983). None of the felids maintains a permanent pair bond (Eltringham 1979). In all but the social cats, the young are cared for mainly by their mothers (McBride 1990).

Infanticide by immigrating males is a regular phenomenon in African lions (Packer and Pusey 1983). This is often followed by synchronized breeding in all the females of a pride (Packer and Pusey 1987). Starvation and predation are major causes of cub mortality among felids. This is especially so in arid regions such as the southern Kalahari, where cub mortality is high in both the lion (Eloff 1973a) and the leopard (Bothma and Le Riche 1984, 1986). In the cheetah, 50% of the cubs die before they are six months old in the southern Kalahari (Labuschagne 1979). Even in the prey-rich, relatively mesic Serengeti, only 4.8% of the cheetah cubs born reach independance (Caro 1994).

Hyaenids are notorious for the apparent, but false hermaphrodism found in adult spotted hyaenas (Gould 1981). This fallacy is generated by the presence of a false scrotum and a highly erectile, peniform clitoris in the adult female (Matthews 1939; Deane 1962; Racey and Skinner 1979). High foetal androgen levels are responsible for the phenomenon (Racey and Skinner 1979). With other factors, these androgon levels eventually lead to the female dominating the clan (Gould 1981). Spotted hyaenas are polyoestrous, with no clearly defined breeding season (Deane 1962; Lindeque and Skinner 1982; Mills 1990; Cooper 1993) The complete sexual cycle lasts for nearly 12 months (Matthews 1939). Spotted hyaena cubs are raised in dens, and are dependent upon their mothers for 14 to 18 months (East and Hofer 1991). Siblicide occurs in same-sex litters (Werdelin and Solounias 1991).

Striped and brown hyaenas are also polyoestrous and breed throughout the year (Skinner and Ilani 1979). Again, the cubs are raised in dens (Kruuk 1976; Mills 1990). In the Kalahari, nomadic brown hyaena males do most of the mating, but the usual clan tenure of a given nomadic male is only 26 months (Owens and Owens 1978; Mills 1990). Allo-parenting of brown hyaena cubs occurs in some parts of the Kalahari (Owens and Owens 1979a). Aardwolves form mated, monogamous pairs which defend a common territory. Nevertheless, a female mated to a less aggressive male will also mate with a neighbouring and more aggressive male. However, she will often mate with her own male too as her period of sexual receptivity lasts several days. This means that a specific male may be cuckolded into rearing another male's cubs while they are confined to the den (Richardson 1987c).

In some of the herpestids, such as the dwarf mongoose of the Taru Desert, only the dominant male and female breed. Subordinate members are suppressed reproductively (Rasa 1987a; Creel et al. 1992). In others, such as the banded mongoose, several females in a pack may breed (Macdonald 1983). Subordinate female dwarf mongooses are suppressed reproductively by a complex of endocrine and behavioural mechanisms. Subordinate males are suppressed mainly by behavioural mechanisms alone (Creel et al. 1992). There is a significant ranking correlation between mating and age in dwarf mongooses. Age has a greater impact on sexual dominance than body mass (Rood 1990). Dwarf mongooses use subordinate helpers to raise the young (Jennions and Macdonald 1994), which are guarded at the den while the rest of the pack forages (Rasa 1987a).

Mating in the slender mongoose is reminiscent of that of the cheetah in that a female in oestrus is followed by several male coalition members (Waser et al. 1994). The yellow mongoose in the arid parts of southern Africa is unique in producing two successive litters, the second oestrus period coinciding with lactation. However, in the arid Karoo region, a second litter is absent when food is scarce (Rasa et al. 1992).The suricate is reproductively active over a long period and females in all stages of reproduction are found in a population (Lynch 1980).

7.4 Area Use, Movements, and Activity

There is considerable interspecific and intraspecific variation among carnivores. Consequently, ecological factors alone are of limited use for predicting spatial organization (Hofer and East 1993a). For solitary female carnivores, food is particularly important. The range must therefore contain sufficient resources for the most severe times. Consequently, there is a correlation between prey abundance and range size. Exclusive ranges are to be expected when food resources are stable and evenly distributed. Range use by males is influenced mainly by food resources, and by receptive females in the breeding season (Macdonald 1983; Sandell 1989). A social system with exclusive ranges is, however, only possible if it is best for all the dominant males. This is especially so when female density is high and the females are evenly distributed (Sandell 1989).

The advantages of knowing intimately the resources available in a particular range may well favour ranges in carnivores that are as large, but no larger than necessary to support a population of them (Macdonald 1983). Almost all carnivore territories are scent-marked to keep intruders out. However, in some social species, such as the wild dog, only the highest ranking pack members scent-mark (Gorman and Trowbridge 1989). Activity levels in carnivores are correlated with the basal rate of metabolism that has to be maintained. In broad terms, and with many exceptions, carnivores from deserts have low basal rates of metabolism (McNab 1989). However, as desert nights can be cold, strategies such as den use may be employed to overcome periods of stress.

Variations in prey size, and the temporal and spatial availability of food greatly affect the spacing system of canids. Small species which feed primarily on invertebrates and fruit, have overlapping ranges with little impact on the foraging success of each other. Among territorial, solitary foragers such as most felids, the distribution of food may determine the size of the territory. In territorial canids, ecological constraints determine whether it is possible for helpers to remain in the territory of the parental pair (Moehlman 1989).

The ranges used by wild dog packs vary and overlap greatly because they are nomadic carnivores (Burrows 1995). The movements of the a pack within the range are related to prey abundance and hunting success. The dominant pair determines these movements (Fuller and Kat 1990). Large range size requirements have today confined most of the world's population of wild dogs to large conservation areas. Among the jackals, range size also seems to be related to food resource availability (e.g. Wyman 1967; Moehlman 1989). In the case of the black-backed jackal of the southern Kalahari and the Namib Desert, range size increases in these more arid regions (Nel 1978; Ferguson et al. 1983; Hiscocks and Perrin 1988).

Felid males generally occupy large, exclusive ranges that overlap with those of several females. The range size of a female is related to the availability of food resources, especially during the lean season (Sandell 1989; Sunquist and Sunquist 1989). The range of the lion in Etosha National Park varies greatly and is clearly related to pride size (Stander 1991). In the southern Kalahari, leopards and cheetahs have range sizes much larger than in mesic areas (Labuschagne 1979; Bothma, unpubl.). Among some cheetahs, body size and age are two crucial factors influencing whether individual cheetahs become territorial or not (Caro and Collins 1987b).

Hyaena territories are extensively scent-marked (Cooper 1990b; Mills 1990; Richardson 1987c, 1990). On the fertile grasslands of East Africa, with their abundant migrating prey resources, spotted hyaenas live in large territorial clans. However, where the prey are sedentary, their territories are small (Cooper 1990b). Territory integrity may be breached by spotted hyaenas in response to the problem of variable resources. They may commute over several territories to the nearest food and water (Hofer and East 1993a-c; Cooper 1990b). In the Serengeti, the critical resource which triggers commuting in spotted hyaenas are the migratory herds of prey (Hofer and East 1993a-c) but ironically in the Savuti Marsh, which dries up for many years at a time, it is the presence of water (Cooper 1990b). In the Namib Desert, water is a critical factor in the maintenance of territories by spotted hyaenas. Their territories are large compared with those of spotted hyaenas in more mesic areas (Tilson and Henschel 1986; Henschel and Skinner 1987; Mills 1990; Hofer and East 1995). When food and water resources are evenly spaced in the Savuti, the territory size of spotted hyaena clans is similar in different places (Cooper 1990b).

Striped hyaenas do not appear to defend a territory (Kruuk 1976). Brown hyaenas in the southern Kalahari have large, overlapping territories, the size of each being related to the amount of food present (Mills 1978a, 1983; Mills and Mills 1982). Brown hyaena ranges are even larger in the Namib Desert than in the southern Kalahari, and these sizes are dictated solely by the available food (Mills 1990). Aardwolf territorial boundaries may be transgressed in the mating season by cuckolding, more aggressive, and promiscuous males (Richardson 1987c).

Range size is dependent upon the food resources and varies between species in the case of many smaller carnivores. A dwarf mongoose territory in the Taru Desert may contain up to 200 termite mounds which are used as overnight dens (Rasa 1989a). Slender mongooses may have home ranges occupied by several males which form a loose association (Waser et al. 1994).

Many of the larger carnivores are nocturnally active, but some hunt mainly by day. In the central parts of South Africa there is a general gradient in 32% of the 33 species of carnivore studied, ranging from a nocturnal and solitary mode of life to an entirely diurnal and gregarious one. Carnivores are considered to be incapable of adapting both to an entirely nocturnal and a gregarious life style (Rautenbach and Nel 1978). Some carnivores, such as the bat-eared fox, will change their activity cycles in response to season (Rautenbach and Nel 1978) and the availability of food (Koop and Velimirov 1982). Aardwolves are also adapted to the activity cycles of their main food resource (Richardson 1987b).

7.5 Feeding Ecology

7.5.1 Food

It is now commonly accepted that all mammalian carnivores primarily hunt or forage. However, most also scavenge whenever food is available (Houston 1979). The diet of large predators is usually opportunistic. To a large extent, diet is determined by what is available (Viljoen and Davis 1973; Eltringham 1979). However, the question involved in diet selection is sometimes not what prey to select, but what to ignore (Stephens and Krebs 1986). Among carnivores, the size of the prey selected generally increases with the size of the predator, or the combined mass of a group of predators. Hence, social predators are able to exploit a wider prey spectrum than solitary ones of similar size (Belsky 1987). A notable exception is the aardwolf, a relatively large predator which is specialized for eating termites (Richardson 1987a,b). Larger carnivores do not only pursue larger prey, but also a wider range of prey sizes (Belsky 1987).

Canids are generally less strictly carnivorous than felids. This makes them greater opportunists and promotes peaceful feeding. This is rare among felids (Kleiman 1967; Kleiman and Eisenberg 1973). The type and size of the food used by canids tends to be correlated with the species' spacing and feeding habits. Small canids which depend upon invertebrates and wild fruits, occupy overlapping ranges and have little impact upon each other's foraging success (Moehlman 1989). The ecological parameters of the environment greatly influence the diet of canids. This is aptly illustrated by the 92% occurrence of plant material, and dependence upon scavenged seal pups as found in the scats of black-backed jackals along the Namib Desert coast (Stuart 1976: Stuart and Shaughnessy 1984). These jackals even catch fish in the tempory lagoons that form after occasional flooding of the mostly dry rivers (Nel and Loutit 1986). Similarly, the Simien fox of the Bale Mountains lives almost exclusively upon an abundant rodent food resource (Sillero-Ziburi and Gottelli 1995).

Felids are generally the most carnivorous of the Carnivora (Kleiman and Eisenberg 1973). They also show a high degree of dependence on vertebrate food (Houston 1979; Sunquist and Sunquist 1989). Large felids are often opportunistic within a given range of alternatives. However, the modal prey mass usually seems to be less than that of the predator's, although this relationship varies greatly in different regions (Sunquist and Sunquist 1989). Like other carnivores, some of the felids scavenge food when available (Houston 1979). Others, such as the cheetah and the caracal, do not scavenge or do so infrequently (Skinner 1979; Caro 1982).

Hyaenids use a wide and varied prey base ranging from large ungulates and carrion, to almost exclusively termites (Richardson 1987a, b; Mills 1990). Differences in prey availability lead to differences in the diet of the spotted hyaena (Kruuk 1966, 1972b). When large ungulates are relatively scarce, as in the southern Kalahari, spotted hyaenas expand their diet to include a wide array of food items, including wild fruit (Mills 1978b, 1990). In the Namib Desert, the diet selected by the spotted hyaena depends upon its foraging range and the availability of prey (Tilson et al. 1980). Striped and brown hyaenas are effective scavengers and poor hunters, and their diet is supplemented by wild fruits when these are available (Bouskila 1984; Mills 1990). The brown hyaena regularly eats ostrich eggs in the southern Kalahari (Mills 1990). Although it is almost entirely a termite eater, cold nights in winter which halt the activity of their main nocturnal prey species, can force the aardwolf to utilize another species of termite which forages mainly during the day in winter (Richardson 1987b,c).

Food resource renewability is of greater importance than prey size in social mongooses (Rasa 1994). Most mongooses eat insects extensively. However, some, such as the dwarf mongoose, the suricate and the yellow mongoose, may eat a wide variety of other foods too (Lynch 1980; Rasa 1989a). A yellow mongoose even kills prey as large as a springhare (Lynch 1980). The Cape grey mongoose is more of a generalist and opportunistic feeder (Cavallini 1992).

7.5.2 Hunting and Foraging

The carnivores vary greatly in their hunting and foraging methods. These differences even exist within populations of the same species, depending upon factors such as the behaviour of the prey, and the social rank of the individual carnivore. Yet some factors are common to several families. These include cooperative hunting (Schaller and Lawther 1969). This strategy has to date been found in canids, felids and hyaenids, (Macdonald 1983), all of which include social species (Bertram 1979). Hunting as a group allows predators to kill prey larger than what they would be able to tackle if they were solitary. However, by hunting alone, some species can also exploit smaller prey (Belsky 1987). Those predators that rely on stalking, hunt mainly at night, while those that run their prey down do so in daylight (Kruuk and Turner 1967).

All species of canid show at least some degree of cooperative hunting (Sandell 1989), although it may not be a general rule. The trend towards cooperative hunting increases among canids as body size increases (Moehlman 1989). Consequently, communal hunting is most ritualized in the African wild dog where it assumes an important role in the maintaince of social cohesion within the pack (Kleiman 1967).

Felids differ widely in their hunting methods. Some even employ complicated stalking techiques. The main aim is to get as close as possible to the intended prey without causing it to flee (Eisenberg and Leyhausen 1972). The cheetah differs from other felids in that it hunts mainly by day (Caro 1994), and runs down its prey when doing so (Rautenbach and Nel 1978). Contrary to the general belief, the lion is capable of sustained bursts of speed over as much as 1 km when chasing prey (McBride 1990). In the Savuti Marsh, the relative absence of large males in many prides, and consequent kleptoparasitism by spotted hyaenas, results in the loss of many kills. However, this usually does not happen before the lions have fed (Cooper 1991). The lion and leopard have flexible hunting techniques which are prey-specific in the southern Kalahari (Eloff 1973a, 1984; Bothma and Le Riche 1989). Despite its aridity, the southern Kalahari provides sufficient hunting cover for leopards (Bothma et al. 1994). Hunger there seems to motivate increased hunting effort (Bothma and Le Riche 1990).

Although there are only four extant species, all of them monotypic, the hyaenids show a wide diversity in hunting and foraging strategies (Mills 1990). In dry, desert regions, mobile prey often follow the sparse rains. This forces spotted hyaenas to range over large areas (Cooper 1990b). In the Serengeti, on the other hand, migratory prey are reached by commuting through the territories of other clans (Hofer and East 1993a-c). Occasionally, members of spotted hyaena clans assemble temporarily in large parties to hunt cooperatively (Macdonald 1983). Spotted hyaenas of the Savuti are adept at kleptoparasitism and often steal kills from lion prides which contain no large, adult male (Cooper 1991). Nevertheless, they hunt their own food most of the time (Cooper 1990a).

Striped and brown hyaenas usually forage alone (Bouskila 1984; Mills 1990). Some brown hyaenas do hunt small prey, however, while others even become habitual killers of a particular type of prey (Mills 1990). The aardwolf forages alone for termites which it licks up with its broad tongue (Bothma and Nel 1980; Richardson 1987c). Herpestids forage in packs or alone (e.g. Rasa 1987a; Lynch 1980). Foraging dwarf mongooses maintain contact by means of individually identifiable calls (Rasa 1986). Some herpestids, such as the yellow mongoose, are silent when foraging, but others like the suricate are quite vocal (Lynch 1980).

7.5.3 Killing

The familiarity of a predator with its prey, and the reaction of the prey to predation, largely determines the killing response. With some prey, a predator must learn to employ a special technique (Eisenberg and Leyhausen 1972). Surplus killing occurs occasionally in a variety of predators in response to environmental pressures other than hunting, and which lead to a concentration of vulnerable prey (Kruuk 1972a). Among the canids of Africa, only the wild dog is without any special behavioural pattern for killing its prey (Kruuk and Turner 1967). Lions in the southern Kalahari approach the potentially lethal gemsbok with care, and employ a special killing technique (Sect. 3.5.2) (Eloff 1973a, 1984). This technique is absent from the Etosha National Park (Eloff 1973a; Stander 1992a,b). Killing rates among the leopards of the southern Kalahari are relatively low (Bothma and Le Riche 1986), as they are elsewhere too (e.g. Schaller 1972). In comparison, the rates of kill of spotted hyaenas are much higher (Mills 1990).

7.5.4 Feeding

Among carnivores, there are significant interactions between the use of the teeth and the type of food eaten. Rather than being used for cutting muscle, the carnassial teeth are essential for processing tough food such as skin, and muscle attached to bone. There is also a negative relationship between the size of the predator and the need for food to be chewed. In carnivores that feed in groups, speed of ingestion may be more important than precision in the use of the teeth. Canids and hyaenids eat more bone than felids (Van Valkenburgh 1996). In some canids and hyaenids, food may be cached for later use (e.g. Macdonald 1976; Malcolm 1980; Mills 1990).

7.5.5 Water Use

Carnivores obtain most of their water from the body fluids of their prey in arid lands (Delany and Happold 1979). These fluids provide appreciable quantities of water which remain the same in summer and winter. The most common carnivores of deserts are canids such as jackals and foxes. Of all the species of Carnivora in the Sahara Desert, only the fennec fox is found often near water (Schmidt-Nielsen 1964), although it can live without it for long periods (Von Gauthier-Pilters 1967). For spotted hyaenas, however, water is important (Skinner and Van Aarde 1981). In the Savuti Marsh, spotted hyaenas transgress territorial boundaries in search of water (Cooper 1990b). In the Etosha National Park and the Namib Desert, each clan of spotted hyaenas is dependent upon a permanent source of water in its terrritory (Gasaway et al. 1991; Tilson and Henschel 1986). Free water is also vital for the striped hyaena (Rieger 1979a), but not for the brown hyaena (Mills 1990).

7.6 Population Ecology

Large carnivores, such as African wild dogs, need large conservation areas if they are to survive (Fanshawe et al. 1991; Burrows 1995). Cub mortality in the African wild dog is high, and 90% never reach maturity (Fanshawe et al. 1991). Much the same is true of large felids such as the lion (Schaller 1972; Eloff 1973a) and the cheetah (Caro 1994), but not so of the spotted and brown hyaena (Mills 1990). Most large carnivores seem to be in dynamic equilibrium with their environment.

7.7 Interactions with Other Carnivores

Interspecific competition is avoided or reduced between most co-existing carnivores, primarily through dietary differences, differential use of habitats (Rautenbach and Nel 1978; Bertram 1979), and activity patterns (Fuller et al. 1989). Current evidence shows no causal relationship between cooperative hunting and group living (Caro 1994). Nevertheless, a clear dominance order appears among co-existing carnivores if they should compete for the same resource. This is especially marked between lions and spotted hyaenas. The balance of power depends upon the size of each group, and the presence or absence of an adult male lion within the pride (e.g. Schaller 1972; Mills 1990; Cooper 1991; Stander 1992b). Of the larger carnivores, the brown hyaena often ranks lowest on the scale of dominance, or is usually close to it (Mills 1990).

7.8
Summary

The 268 species of living Carnivora known to the world today show a high degree of diversity, but they also share some marked similarities. The carnivores occupy a wide range of habitats, and some do particularly well in arid environments. The larger carnivores, however, require extensive conservation areas if they are to avoid conflict with man. In this regard, arid lands could well be the ultimate agent of conservation for the survival of these carnivores because the mesic regions are usually most heavily populated by man. Consequently, large conservation areas are difficult to proclaim and maintain there. In arid areas it may be less of a problem. However, for effective conservation, even in arid lands, a reliable and extensive scientific database will have to be built up for many of the species of Carnivora for which it is often still lacking today, particularly in such regions.

References

Aiken B (1987) Nightstalk. Afropix, Jersey

Al-Khalili AD (1993) Ecological review and the distribution of Blanford's fox *Vulpes cana* Blanford, 1877 (Mammalia : Carnivora : Canidae). Fauna Saudi Arabia 13:390-396

Anderson MD, Richardson PRK, Woodall PF (1992) Functional analysis of the feeding apparatus and digestive tract of the aardwolf *Proteles cristatus*. J Zool (Lond) 228:423-434

Anholt BR, Ludwig D, Rasmussen JB (1987) Optimal pursuit times: how long should predators pursue their prey? Theor Popul Biol 31:453-464

Anonymous (1996a) Lumping leopards; should 27 subspecies be reduced to eight? Cat News 25:9

Anonymous (1996b) Lions in Addis Ababa arouse speculation about extinct races. Cat News 25:2-3

Anyonge W (1996) Microwear on canines and killing behavior in large carnivores: saber function in *Smilodon fatalis*. J Mammal 77(4):1059-1067

Apps PJ, Viljoen HW, Richardson PRK, Pretorius V (1989) Volatile components of anal gland secretion of the aardwolf (*Proteles cristatus*). J Chem Ecol 15(5):1681-1688

Asa CS, Mech LD, Seal US (1985) The use of urine, faeces, and anal-gland secretions in scent-marking by a captive wolf (*Canis lupus*) pack. Anim Behav 33(3):1034-1036

Avenant NL (1993) The caracal, *Felis caracal caracal* Schreber, 1776, as predator in the West Coast National Park. MSc Thesis, University of Stellenbosch, Stellenbosch

Avery G, Avery DM, Braine S, Loutit R (1987) Prey of coastal black-backed jackal *Canis mesomelas* (Mammalia : Canidae) in the Skeleton Coast Park, Namibia. J Zool (Lond) 213:81-94

Bailey TN (1993) The African leopard. Columbia University Press, New York

Baker W (1982) Loving leopards. Swara 5(4):26-27

Bearder SK (1977) Feeding habits of spotted hyaenas in a woodland habitat. East Afr Wildl J 15:263-280

Bearder SK, Randall RM (1978) The use of fecal marking sites by hyaenas and civets. Carnivore 1(2):32-48

Belsky AJ (1987) A flexible body mass in social carnivores. Am Nat 129(5):755-760

Beltràn JF, Delibes M (1994) Environmental determinants of circadian activity of free-ranging Iberian lynxes. J Mammal 75(2):382-393

Berry H (1983) First catch your lion. Rössing Magazine, April 1983:1-7

Berry MPS (1981) Stomach contents of bat-eared foxes, *Otocyon megalotis*, from the northern Transvaal . S Afr J Wildl Res 11:28-30

Bertram BCR (1974) Radio-tracking leopards in the Serengeti. Afr Wildl Lead Found News 9(2):7-10

Bertram BCR (1979) Serengeti predators and their social systems. In: Sinclair ARE, Norton-Griffiths M (eds) Serengeti, dynamics of an ecosystem. University of Chicago Press, Chicago, pp 221-248

Beynon P, Rasa OAE (1989) Do dwarf mongooses have a language?: warning vocalisations transmit complex information. S Afr J Sci 85: 447-450

Biquard S, Urios V, Boug A, Vila C, Castroviejo J, Nader I (1994) Fishes as diet of a wolf (*Canis lupus arabs*) in Saudi Arabia. Mammalia 58(3): 492-494

Boggs LP, McNutt JW (1994) Running wild. Africa 2(5):28-36

Bothma JduP (1966) Food of the silver fox *Vulpes chama*. Zool Afr 2(2):205-210

Bothma JduP (1971a) Control and ecology of the black-backed jackal *Canis mesomelas* in the Transvaal. Zool Afric 6(2):187-193

Bothma JduP (1971b) Notes on movement by the black-backed jackal and the aardwolf in the western Transvaal. Zool Afric 6(2):205-207

Bothma JduP (1971c) Food of *Canis mesomelas* in South Africa. Zool Afr 6(2):195-203.

Bothma JduP (1971d) Food habits of some Carnivora (Mammalia) from southern Africa. Ann Transvaal Mus 27(2):15-26

Bothma JduP, Knight MH, Le Riche EAN, Van Hensbergen HJ (In press) Range size of southtern Kalahari leopards. S Afr J Wildl Res 27

Bothma JduP, Le Riche EAN (1984) Aspects of the ecology and the behaviour of the leopard *Panthera pardus* in the Kalahari Desert. Koedoe Suppl 27:259-279

Bothma JduP, Le Riche EAN (1986) Prey preference and hunting efficiency of the Kalahari Desert leopard. In : Miller SD, Everett DD (eds) Cats of the world: biology, conservation and management. National Wildlife Federation, Washington, DC, pp 389-414

Bothma JduP, Le Riche EAN (1989) Evidence of a flexible hunting technique in Kalahari leopards. S Afr J Wildl Res 19(2):57-60

Bothma JduP, Le Riche EAN (1990) The influence of increasing hunger on the hunting behaviour of southern Kalahari leopards. J Arid Environ 18:79-84

Bothma JduP, Le Riche EAN (1993) Disturbance bias when tracking Kalahari leopards *Panthera pardus* by spoor. Koedoe 36(2):109-112

Bothma JduP, Le Riche EAN (1994a) Scat analysis and aspects of defecation in northern Cape leopards. S Afr J Wildl Res 24(1,2): 21-25

Bothma JduP, Le Riche EAN (1994b) The relationship between minimum air temperature and daily distance moved by Kalahari leopards. S Afr J Wildl Res 24(1,2):18-20

Bothma JduP, Le Riche EAN (1995) Evidence of the use of rubbing, scent-marking and scratching posts by Kalahari leopards. J Arid Environ 29:511-517

Bothma JduP, Nel JAJ (1980) Winter food and foraging behaviour of the aardwolf *Proteles cristatus* in the Namib-Naukluft Park. Madoqua 12(3):141-149

Bothma JduP, Nel JAJ, Macdonald A (1984) Food niche separation between four sympatric Namib Desert carnivores. J Zool (Lond) 202:327-340

Bothma JduP, Van Rooyen N, Theron GK, Le Riche EAN (1994) Quantifying woody plants as hunting cover for southern Kalahari leopards. J Arid Environ 26:273-280

Bothma JduP, Van Rooyen N, Le Riche EAN (1997) Multivariate analysis of the hunting tactics of Kalahari leopards. Koedoe 40(1):41-56

Bouskila Y (1984) The foraging groups of the striped hyaena (*Hyaena hyaena syriaca*). Carnivore 7:2-12

Bridgeford PA (1985) Unusual diet of the lion *Panthera leo* in the Skeleton Coast Park. Madoqua 14(2):187-188

Brocklesby DW, Vidler BO (1965) Observations on the behaviour of young spotted hyaena (*Crocuta crocuta*) in the burrow. East Afr Wildl J 3:122-123

Bronner GN (1992) Burrow system characteristics of seven small mammal species (Mammalia: Insectivora; Rodentia; Carnivora). Koedoe 35(1):125-128

Bryden BR (1978) The biology of the African lion (*Panthera leo*) (Linn. 1758) in the Kruger National Park. MSc Thesis, University of Pretoria, Pretoria

Burrows R (1995) Demographic changes and social consequences in wild dogs. In : Sinclair ARE, Arcese P (eds) Serengeti II: dynamics, management and conservation of an ecosystem. University of Chicago Press, Chicago, pp 400-420

Caraco T, Wolf LL (1975) Ecological determinants of group sizes of foraging lions. Am Nat 109(967):343-352

Carlstead K (1991) Husbandry of the fennec fox *Fennecus zerda:* environmental conditions influencing stereotypic behaviour. Int Zoo Yearb 30:202-207

Caro TM (1982) A record of cheetah scavenging in the Serengeti. Afr J Ecol 20:213-214

Caro TM (1994) Cheetahs of the Serengeti Plains. University of Chicago Press, Chicago

Caro TM, Collins DA (1987a) Ecological characteristics of territories of male cheetahs (*Acinonyx jubatus*). J Zool (Lond) 211:89-105
Caro TM, Collins DA (1987b) Male cheetah social organization and territoriality. Ethology 74:52-64
Caro TM, Laurenson MK (1994) Ecological and genetic factors in conservation : a cautionary tale. Science 263:485-486
Cavallini P (1992) *Herpestes pulverulentus*. Mammal Species 409:1-4
Cavallini P, Nel JAJ (1995) Comparative behaviour and ecology of two sympatric mongoose species (*Cynictis penicillata* and *Galleria pulverulenta*). S Afr J Wildl Res 30(2):46-49
Cavallini P, Serafini P (1995) Winter diet of the small Indian mongoose, *Herpestes auropunctatus*, on an Adriatic Island. J Mammal 76(2):569-574
Chellam R, Johnsingh AJT (1993) Management of Asiatic lions in the Gir forest, India. Symp Zool Soc (Lond) 65:409-424
Child G, Robbel H (1975) Drowning of lechwe by spotted hyaena. Mammalia 39(4):705
Childes SL (1988) The past history, present status and distribution of the hunting dog *Lycaon pictus* in Zimbabwe. Biol Conserv 44:301-316
Cloudsley-Thompson JL (1982) Arid lands in perspective - a biologist's view. Ann Arid Zone 21(2):98-103
Cloudsley-Thompson JL (1993) The adaptational diversity of desert biota. Environ Conserv 20(3):227-231
Cloudsley-Thompson JL (1996) Biotic interactions in arid lands. Adaptations of desert organisms. Springer, Berlin Heidelberg New York
Cooper SM (1990a) The hunting behaviour of spotted hyaenas (*Crocuta crocuta*) in a region containing both sedentary and migratory populations of herbivores. Afr J Ecol 28:131-141
Cooper SM (1990b) Clan sizes of spotted hyaenas in the Savuti region of the Chobe National Park, Botswana. Botswana Notes Rec 21:121-133
Cooper SM (1991) Optimal hunting group size: the need for lions to defend their kills against loss to spotted hyaenas. Afr J Ecol 29:130-136
Cooper SM (1993) Denning behaviour of spotted hyaenas (*Crocuta crocuta*) in Botswana. Afr J Ecol 31:178-180
Corbett LK, Newsome AE (1987) The feeding ecology of the dingo. III: Dietary relationships with widely fluctuating prey populations in arid Australia: an hypothesis of alternation of predation. Oecologia 74:215-227
Creel S, Creel N, Weldt DE, Monfort SL (1992) Behavioural and endocrine mechanisms of reproductive suppression in Serengeti dwarf mongooses. Anim Behav 43:231-245
Curio E (1976)The ethology of predation. Zoophysiology and ecology 7. Springer, Berlin Heidelberg New York
De Vos V (1970) Pseudopregnancy in the black-backed jackal (*Canis mesomelas* Schreber). J S Afr Vet Med Assoc 40(4):381-383
Dean WRJ, Milton SJ (1991) Disturbances in semi-arid shrubland and arid grassland in the Karoo, South Africa: mammal diggings as germination sites. Afr J Ecol 29:11-16
Deane NN (1962) The spotted hyaena *Crocuta crocuta*. Lammergeyer 2:26-44
Delany MJ, Happold DCD (1979) Ecology of African mammals. Longman, London
Dorst J, Dandelot P (1970) A field guide to to the larger mammals of Africa. Collins, London
Dragesco-Joffé A (1993) La vie sauvage au Sahara. Delach aux Niestle, Lausanne
Dreyer HVA, Nel JAJ (1990) Feeding site selection by black-backed jackals on the Namib Desert coast. J Arid Environ 19:217-224
Earlé RA (1981) Aspects of the social and feeding behaviour of the yellow mongoose *Cynictis penicillata* (G.Cuvier). Mammalia 45(2):143-152
East ML, Hofer H (1991) Loud calling in a female-dominated mammalian society. II. Behavioural contexts and functions of whooping of spotted hyaenas, *Crocuta crocuta*. Anim Behav 42:651-669
East ML, Hofer H, Turk A (1989) Functions of birth dens in spotted hyaenas. J Zool (Lond) 219:690-697

Eaton RL (1970a) Group interactions, spacing and territoriality in cheetahs. Z Tierpsychol 27:481-491

Eaton RL (1970b) Hunting behavior of the cheetah. J Wildl Manage 34(1):56-67

Eaton RL (1970c) The predatory sequence, with emphasis on killing behavior and its ontogeny, in the cheetah (*Acinonyx jubatus* Schreber). Z Tierpsychol 27:492-504

Eaton RL (1976a) The brown hyaena : a review of biology, status and conservation. Mammalia 40(3):377-399

Eaton RL (1976b) Support of the hypothesis that aardwolves mimic hyaenas. Mammalia 40(2):342-343

Eisenberg JF (1970) A splendid predator does its own thing untroubled by man. Smithsonian 1(6):48-53

Eisenberg JF, Leyhausen P (1972) The phylogenesis of predatory behavior in mammals. Z Tierpsychol 30:59-93

Eisenberg JF, Lockhart M (1972) An ecological reconnaissance of Wilpattu National Park, Ceylon. Smithsonian Contrib Zool 101:1-118

Eloff FC (1964) On the predatory habits of lions and hyaenas. Koedoe 7:105-112

Eloff FC (1973a) Ecology and behavior of the Kalahari lion *Panthera leo vernayi* (Roberts). In: Eaton RL (ed) The world's cats. World Wildlife Safari, Winston, pp 90-126

Eloff FC (1973b) Water use by the Kalahari lion *Panthera leo vernayi.* Koedoe 16:149-154

Eloff FC (1973c) Lion predation in the Kalahari Gemsbok National Park. J S Afr Wildl Manage Assoc 3(2):59-63

Eloff FC (1980) Cub mortality in the Kalahari lion *Panthera leo vernayi* (Roberts, 1948). Koedoe 23:163-170

Eloff FC (1984) Food ecology of the Kalahari lion *Panthera leo vernayi.* Koedoe Suppl 27:249-258

Eltringham SK (1979) The ecology and conservation of large African mammals. Macmillan, Lond.

Estes RD, Goddard JD (1967) Prey selection and hunting behaviour of the African wild dog. J Wildl Manage 31(1):52-70

Ewer RF (1973) The carnivores. Cornell University Press, London

Fagotto F (1985) The lion in Somalia. Mammalia 49(4):587-588

Fanshawe JH, Frame LH, Ginsberg JR (1991) The wild dog - Africa's vanishing carnivore. Oryx 25(3):137-146

Ferguson JWH (1978) Social interactions of black-backed jackals *Canis mesomelas* in the Kalahari Gemsbok National Park. Koedoe 21:151-162

Ferguson JWH, Nel JAJ, De Wet MJ (1983) Social organization and movement patterns of black-backed jackals *Canis mesomelas* in South Africa. J Zool (Lond) 199:487-502

Ferguson JWH, Galpin JS, De Wet MJ (1988) Factors affecting the activity patterns of black-backed jackals *Canis mesomelas.* J Zool (Lond) 214:55-59

Fey V (1964) The diet of leopards. Afr Wildl 18:105-108

FitzGibbon CD, Fanshawe JH (1989) The condition and age of Thomson's gazelles killed by cheetahs and wild dogs. J Zool (Lond) 218:99-107

Frame LH, Malcolm JR, Frame GW, Van Lawick H (1979) Social organization of African wild dogs (*Lycaon pictus*) on the Serengeti Plains, Tanzania 1967-1978. Z Tierpsychol 50(3):225-249

Frame GW, Frame LH (1981) Swift and enduring: cheetahs and wild dogs of Serengeti. Dutton, New York

Frame GW (1992) First record of the king cheetah in West Africa. Cat News 17:2-3

Frank LG (1986a) Social organization of the spotted hyaena (*Crocuta crocuta*). I. Demography. Anim Behav 34:1500-1509

Frank LG (1986b) Social organization of the spotted hyaena (*Crocuta crocuta*). II. Dominance and reproduction. Anim Behav 34:1510-1527

Frank LG, Holekamp KE, Smale L (1995) Dominance, demography, and reproductive success of female spotted hyaenas. In: Sinclair ARE , Arcese P (eds) Serengeti II: dynamics, management and conservation of an ecosystem. University of Chicago Press, Chicago, pp 364-384

Fuller H, Darby HC (1958) The university atlas. Philip, London

Fuller TK, Biknevicius AR, Kat PW (1988) Home range of an African wildcat, *Felis silvestris* (Schreber), near Elmenteita, Kenya. Z Saugetierkd 53: 380-381

Fuller TK, Biknevicius AR, Kat PW, Van Valkenburgh B, Wayne RK (1989) The ecology of three sympatric jackal species in the Rift Valley of Kenya. Afr J Ecol 27:313-323

Fuller TK, Kat PW (1990) Movements, activity and prey relationships of African wild dogs (*Lycaon pictus*) near Aitong, southwestern Kenya. Afr J Ecol 28:330-350

Fuller TK, Kat PW (1993) Hunting success of African wild dogs in southwestern Kenya. J Mammal 74(2):464-467

Fuller TK, Nicholls TH, Kat PW (1995) Prey and estimated food consumption of African wild dogs in Kenya. S Afr J Wildl Res 25(3):106-110

Garland T (1983) The relation between maximal running speed and body mass in terrestrial mammals. J Zool (Lond) 199:157-170

Gasaway WC, Mossestad KT, Stander PE (1989) Demography of spotted hyaenas in an arid savanna, Etosha National Park, South West Africa/Namibia. Madoqua 16(2):121-127

Gasaway WC, Mossestad KT, Stander PE (1991) Food acquisition by spotted hyaenas in Etosha National Park, Namibia: predation versus scavenging. Afr J Ecol 29:64-75

Gittleman JL (1989) Carnivore group living: comparative trends. In: Gittleman (ed) Carnivore behavior, ecology and evolution. Chapman and Hall, London, pp 183-207

Golightly RT, Ohmart RD (1984) Water economy of two desert canids: coyote and kit fox. J Mammal 65(1):51-58

Gorman ML (1976) A mechanism for individual recognition by odour in *Herpestes auropunctatus*. Anim Behav 24:141-145

Gorman ML, Mills MGL (1984) Scent-marking strategies in hyaenas (Mammalia). J Zool (Lond) 202:535-547

Gorman ML, Trowbridge BJ (1989) The role of odor in the social lives of carnivores. In: Gittleman JL (ed) Carnivore behavior, ecology and evolution. Chapman and Hall, London, pp 57-88

Gosling LM (1982) A reassessment of the function of scent marking in territories. Z Tierpsychol 60:89-118

Goss RA (1986) Maberly's mammals of southern Africa. Delta, Johannesburg

Gould SJ (1981) Hyaena myths and realities. Nat Hist 90(2):16-20

Green B, Anderson J, Whateley T (1984) Water and sodium turnover and estimated food consumption in free-living lions (*Panthera leo*) and spotted hyaenas (*Crocuta crocuta*). J Mammal 65:(4):593-599

Grobler JH, Wilson VJ (1972) Food of the leopard *Panthera pardus* (Linn.) in the Rhodes Matopos National Park, Rhodesia, as determined by faecal analysis. Arnold Rhod 5(35):1-9

Grobler JH, Hall-Martin A, Walker C (1984) Predators of southern Africa. Macmillan, Johannesbg.

Grzimek BB (1975) The African wild dog. In: Grzimek B (ed) Grzimek's animal life encyclopedia. Van Nostrand Reinhold, New York, pp 254-264

Hamilton PH (1976) The movements of leopards in Tsavo National Park, Kenya, as determined by radio-tracking. MSc Thesis, University of Nairobi, Nairobi

Hanby JP, Bygott JD (1979) Population changes in lions and other predators. In: Sinclair ARE, Norton-Griffiths M (eds) Serengeti, dynamics of an ecosystem. University of Chicago Press, Chicago, pp 249-262

Hanby JP, Bygott JD, Packer C (1995) Ecology, demography and behavior of lions in two contrasting habitats: Ngorongoro Crater and the Serengeti Plains. In: Sinclair ARE, Arcese P (eds) Serengeti II: dynamics, management and conservation of an ecosystem. University of Chicago Press, Chicago, pp 315-331

Henschel JR (1986) The socio-ecology of a spotted hyaena *Crocuta crocuta* clan in the Kruger National Park. DSc Dissertation, University of Pretoria, Pretoria

Henschel JR, Skinner JD (1987) Social relationships and dispersal patterns in a clan of spotted hyaenas *Crocuta crocuta* in the Kruger National Park. S Afr J Zool 22(1):18-24

Henschel JR, Skinner JD (1990) The diet of the spotted hyaenas *Crocuta crocuta* in the Kruger National Park. Afr J Ecol 28(1):69-82

Henschel JR, Skinner JD (1991) Territorial behaviour by a clan of spotted hyaenas *Crocuta crocuta*. Ethology 88:223-235

Henschel JR, Tilson RL (1988) How much does a spotted hyaena eat? Perspective from the Namib Desert. Afr J Ecol 26:247-255

Herzig-Straschil B (1977) Notes on the feeding habits of the yellow mongoose *Cynictis penicillata*. Zool Afr 12(1):225-256

Hes L (1991) The leopards of Londolozi. Struik, Cape Town

Hill A (1980) Hyaena provisioning of juvenile offspring at the den. Mammalia 44(4):594-595

Hills DM, Smithers RHN (1980) The "King Cheetah". Arnold Rhod 9(1):1-23

Hines CJH (1990) Past and present distribution and status of the wild dog *Lycaon pictus* in Namibia. Madoqua 17(1):31-36

Hiscocks K, Perrin MR (1987) Feeding observations and diet of black-backed jackals in an arid coastal environment. S Afr J Wildl Res 17(2):55-58

Hiscocks K, Perrin MR (1988) Home range movements of black-backed jackals at Cape Cross Seal Reserve, Namibia. S Afr J Wildl Res 18(3):97-100

Hiscocks K, Perrin MR (1991) Den selection and use by dwarf mongooses and banded mongooses in South Africa. S Afr J Wildl Res 21(4):119-122

Hofer H, East ML (1993a) The commuting system of Serengeti spotted hyaenas: how a predator copes with migratory prey I. Social organization. Anim Behav 46:547-557

Hofer H, East ML (1993b) The commuting system of Serengeti spotted hyaenas: how a predator copes with migratory prey. III. Attendance and maternal care. Anim Behav 46:575-589

Hofer H, East ML (1993c) The commuting system of Serengeti spotted hyaenas: how a predator copes with migratory prey. II. Intrusion pressure and commuter's space use. Anim Behav 46:559-574

Hofer H, East ML (1995) Population dynamics, population size, and the community system of Serengeti spotted hyaenas. In: Sinclair ARE, Arcese P (eds) Serengeti II: dynamics, management and conservation of an ecosystem. University of Chicago Press, Chicago, pp 332-363

Holekamp KE, Cooper SM, Katona CI, Berry NA, Frank LG, Smale L (1997) Patterns of association among female spotted hyaenas (*Crocuta crocuta*). J Mammal 78(1):55-64

Houston DC (1979) The adaptations of scavengers. In: Sinclair ARE, Norton-Griffiths M (eds) Serengeti, dynamics of an ecosystem. University of Chicago Press, Chicago, pp 263-286

Hunter LTB, Skinner JD (1995) Cannibalism in male cheetahs. Cat News 23:13-14

Hunter ML (1996) Fundamentals of conservation biology. Blackwell, Cambridge (U.S.A.)

Ilani G (1990) The spotted ambassadors of a vanishing world. Israel Al 31:16-24

Jaeger MK, Pandit RK, Haque ME (1996) Seasonal differences in territorial behavior by golden jackals in Bangladesh: howling versus confrontation. J Mammal 77(3):768-775

Jennions MD, Macdonald DW (1994) Cooperative breeding in mammals. Tree 9(3):89-93

Johnsingh AJT (1982) Reproductive and social behaviour of the dhole, *Cuon alpinus* (Canidae). J Zool (Lond) 198:443-463

Johnsingh AJT, Chellam R (1991) India's last lions. Zoogoer Magazine, September-October: 16-20

Johnson RP (1973) Scent marking in mammals. Anim Behav 21:521-535

Johnston HH (1884) The river Congo from its mouth to Bolobo. Low, London

Karanth KN, Sunquist ME (1995) Prey selection by tiger, leopard and dhole in tropical forests. J Anim Ecol 64:439-450

Khidas K (1990) Contribution à la connaissance du chacal doré. Facteurs modulant l'orginisation sociale et territoriale de la sous-espèce algérienne (*Canis aureus algirensis* Wagner 1841). Mammalia 54(3):361-375

Kieser JA (1995) Gnathomandibular morphology and character displacement in the bat-eared fox. J Mammal 76(2):542-550

Kleiman DG (1966) Scent marking in the Canidae. Symp Zool Soc (Lond) 18:176-177

Kleiman DG (1967) Some aspects of social behaviour in the Canidae. Am Zool 7:365-372

Kleiman DG (1968) Reproduction in the Canidae. Int Zoo Yearb 8: 3-8

Kleiman DG, Eisenberg JF (1973) Comparisons of canid and felid social systems from an evolutionary perspective. Anim Behav 21: 637-659

Koop K, Velimirov B (1982) Field observations on activity and feeding of bat-eared foxes (*Otocyon megalotis*) at Nxai Pan, Botswana. Afr J Ecol 20:23-27

Kruger JE (1988) Interrelationships between the larger carnivores of the Klaserie Private Nature Reserve, with special reference to the leopard *Panthera pardus* (Linnaeus, 1758) and the cheetah *Acinonyx jubatus* (Schreber, 1775). MSc Thesis, University of Pretoria, Pretoria

Kruuk H (1966) Clan system and feeding habits of spotted hyaenas (*Crocuta crocuta* Erxleben). Nature 209:1257-1258

Kruuk H (1970) Interactions between populations of spotted hyaenas (*Crocuta crocuta* Erxleben) and their prey species. In: Watson A (ed) Animal populations in relation to their food resources. Blackwell, Oxford, pp 359-374

Kruuk H (1972a) Surplus killing by carnivores. J Zool (Lond) 166:233-244

Kruuk H (1972b) The spotted hyaena. A study of predation and social behaviour. University of Chicago Press, Chicago

Kruuk H (1972c) The aardwolf (*Proteles cristatus* Sparrman 1783) as predator of termites. East Afr Wildl J 10:211-227

Kruuk H (1976) Feeding and social behaviour of the striped hyaena (*Hyaena vulgaris* Desmarest). East Afr Wildl J 14:91-111

Kruuk H (1986) Interactions between Felidae and their prey species: a review. In: Miller SD, Everett DD (eds) Cats of the world: biology, conservation and management. National Wildlife Federation, Washington, DC, pp 353-374

Kruuk H, Mills MGL (1983) Notes on food and foraging of the honey-badger *Mellivora capensis* in the Kalahari Gemsbok National Park. Koedoe 26:153-157

Kruuk H, Turner M (1967) Comparative notes on predation by lion, leopard, cheetah and wild dog in the Serengeti area, East Africa. Mammalia 31(1):1-27

Kühme W (1965) Communal food distribution and division of labour in African hunting dogs. Nature 205(4970):443-444

Labuschagne W (1979) Bio-ekologiese en gedragstudie van die jagluiperd *Acinonyx jubatus jubatus* (Schreber, 1775). MSc Thesis, University of Pretoria, Pretoria

Lamprecht J (1979) Field observations on the behaviour and social system of the bat-eared fox (*Otocyon megalotis* Desmarest). Z Tierpsychol 49:260-284

Laurenson MK (1995) Implications of high offspring mortality for cheetah population dynamics.In: Sinclair ARE, Arcese P (eds) Serengeti II: dynamics, management and conservation of an ecosystem. University of Chicago Press, Chicago, pp 385-399

Le Roux PG, Skinner JD (1989) A note on the ecology of the leopard (*Panthera pardus Linnaeus*) in the Londolozi Game Reserve, South Africa. Afr J Ecol 27:167-171

Leyhausen P (1979) Cat behaviour. Garland Press, New York

Lindburg D (1989) When cheetahs are kings. Zoonoz 62(3):5-10

Lindeque M, Skinner JD (1982) Aseasonal breeding in the spotted hyaena (*Crocuta crocuta* Erxleben), in southern Africa. Afr J Ecol 20:271-278

Lindsay IM, Macdonald DW (1986) Behaviour and ecology of Rüppell's fox, *Vulpes rüppelli*, in Oman. Mammalia 50(4):461-474

Lombaard LJ (1971) Age determination and growth curves in the black-backed jackal, *Canis mesomelas* Schreber 1775 (Carnivora: Canidae). Ann Transvaal Mus 27(7):135-181

Lovegrove B (1993) The living deserts of southern Africa. Fernwood Press, Vlaeberg

Lynch CD (1980) Ecology of the suricate *Suricata suricatta* and yellow mongoose *Cynictis penicillata* with special reference to their reproduction. Mem Nat Mus Bloemfontein 14:1-145

Macdonald DW (1976) Food caching by the red fox and other carnivores. Z Tierpsychol 42:170-185

Macdonald DW (1978) Observations on the behaviour and ecology of the striped hyaena, *Hyaena hyaena* in Israel. Isr J Zool 27:189-198

Macdonald DW (1983) The ecology of carnivore social behaviour. Nature 301:379-384.

Mackie AJ, Nel JAJ (1989) Habitat selection, home range use, and group size of bat-eared foxes in the Orange Free State. S Afr J Wildl Res 19(4):135-139

Maddock AH, Mills MGL (1994) Population characteristics of African wild dogs *Lycaon pictus* in the Eastern Transvaal Lowveld, South Africa as revealed through photographic records. Biol Conserv 67:57-62

Main M (1987) Kalahari. Southern Book Publishers, Johannesburg

Malcolm JR (1980) Food caching by African wild dogs (*Lycaon pictus*). J Mammal 61(4):743-744

Malcolm JR (1986) Socio-ecology of bat-eared foxes (*Otocyon megalotis*). J Zool (Lond) [A] 208:457-467

Malcolm JR, Marten K (1982) Natural selection and the communal rearing of pups in African wild dogs (*Lycaon pictus*). Behav Ecol Sociobiol 10:1-13

Maloiy GMO, Kamau JMZ, Shkolnik A, Meir M, Arieli R (1982) Thermoregulation and metabolism in a small desert carnivore : the fennec fox (*Fennecus zerda*) (Mammalia). J Zool (Lond) 198:279-291

Marlow BJ (1993) Predation by the ratel *Mellivora capensis* on chicks of the white-backed vulture *Gyps africanus*. S Afr J Wildl Res 13:24

Martin LD (1989) Fossil history of terrestrial Carnivora. In: Gittleman JL (ed) Carnivore behavior, ecology and evolution. Chapman and Hall, London, pp 536-568

Matthews LH (1939) Reproduction in the spotted hyaena, *Crocuta crocuta* (Erxleben). Philos Trans R Soc Lond [Biol] 230(565):1-78

Mazak V (1964) Note on the lion's mane. Z Saugetierkd 29:124-127

Mazak V (1970) The Barbary lion, *Panthera leo leo* (Linnaeus, 1758); some systematic notes, and an interim list of the specimens preserved in European museums. Z Saugetierkd 35:34-45

McBride C (1990) Liontide. Jonathan Ball, Johannesburg

McNab B (1989) Basal rate of metabolism, body size, and food habits in the Order Carnivora. In: Gittleman JL (ed) Carnivore behavior, ecology and evolution. Chapman and Hall, London, pp 335-354

McShane TO, Grettenberger JF (1984) Food of the golden jackal (*Canis aureus*) in central Niger. Afr J Ecol 22:49-53

Mech LD (1975) Hunting behavior in two similar species of social canids. In: Fox MW (ed) The wild canids. Van Nostrand Reinhold, New York, pp 363-369

Melton DA, Berry HH, Berry CU, Joubert SM (1987) Aspects of the blood chemistry of wild lions, *Panthera leo*. S Afr J Zool 22(1):40-44

Merola M (1994) A reassessment of homozygosity and the case for inbreeding in cheetah *Acinonyx jubatus*: implications for conservation. Conserv Biol 8(3):961-971

Mills MGL (1977) Foraging behaviour of the brown hyaena (*Hyaena brunnea* Thunberg, 1820) in the southern Kalahari. Z Tierpsychol 48:113-141

Mills MGL (1978a) The comparative socio-ecology of the Hyaenidae. Carnivore 1(1):1-7

Mills MGL (1978b) The diet of the brown hyaena *Hyaena brunnea* in the southern Kalahari. Koedoe 21:125-149

Mills MGL (1982a) *Hyaena brunnea*. Mammal Species 194:1-5

Mills MGL (1982b) The mating system of the brown hyaena, *Hyaena brunnea* in the southern Kalahari. Behav Ecol Sociobiol 10:131-136

Mills MGL (1982c) Factors affecting group size and territory size of the brown hyaena, *Hyaena brunnea* in the southern Kalahari. J Zool (Lond) 198:39-51

Mills MGL (1982d) Notes on age determination, growth and measurements of brown hyaenas *Hyaena brunnea* from the Kalahari Gemsbok National Park. Koedoe 25:55-61

Mills MGL (1983) Behavioural mechanisms in territory and group maintenance of the brown hyaena, *Hyaena brunnea*, in the southern Kalahari. Anim Behav 31:503-510

Mills MGL (1989) The comparative behavioral ecology of hyenas: the importance of diet and food dispersion. In: Gittleman JL (ed) Carnivore behavior, ecology and evolution. Chapman and Hall, London, pp 125-142

Mills MGL (1990) Kalahari hyaenas. Unwin Hyman, London

Mills MGL, Biggs HC (1993) Prey apportionment and related ecological relationships between large carnivores in Kruger National Park. Symp Zool Soc (Lond) 65:253-268

Mills MGL, Gorman ML (1987) The scent-marking behaviour of the spotted hyaena *Crocuta crocuta* in the southern Kalahari. J Zool (Lond) 212:483-497

Mills MGL, Mills MEJ (1977) An analysis of bones collected at hyaena breeding dens in the Kalahari Gemsbok National Park. Ann Transvaal Mus 30(14):145-159

Mills MGL, Mills MEJ (1982) Factors affecting the movement patterns of brown hyaenas, *Hyaena brunnea*, in the southern Kalahari. S Afr J Wildl Res 12(4):111-117

Mills MGL, Shenk TM (1992) Predator-prey relationships : the impact of lion predation on wildebeest and zebra populations. J Anim Ecol 61:693-702

Mills MGL, Wolff P, Le Riche EAN, Meyer IJ (1978) Some population characteristics of the lion *Panthera leo* in the Kalahari Gemsbok National Park. Koedoe 21:163-171

Mills MGL, Nel JAJ, Bothma JduP (1984) Notes on some smaller carnivores from the Kalahari Gemsbok National Park. Koedoe Suppl 27:221-227

Mitchell BL, Shenton JB, Uys JCM (1965) Predation on large mammals in the Kafue National Park, Zambia. Zool Afr 1(2): 297-318

Moehlman PD (1979) Jackal helpers and pup survival. Nature 277:382-383

Moehlman PD (1989) Intraspecific variation in canid social systems. In: Gittleman JL (ed) Carnivore behavior, ecology and evolution. Chapman and Hall, London, pp 143-163

Moolman LC (1986) Aspekte van die ekologie en gedrag van die rooikat *Felis caracal* Schreber, 1776, in die Bergkwagga Nasionale Park en op die omliggende plase. MScThesis, University of Pretoria, Pretoria

Moran G, Sorenson L (1986) Scent marking behavior in a captive group of meerkats (*Suricata suricatta*). J Mammal 67(1):120-132

Morris PA, Malcolm JR (1977) The Simien fox in the Bale Mountains. Oryx 14:151-160

Muckenhirn NA, Eisenberg JF (1973) Home ranges and predation of the Ceylon leopard (*Panthera pardus fusca*) In: Eaton RL (ed) The world's cats. World Wildlife Safari, Winston, pp 142-175

Müller-Using D (1975) Jackals. In: Grzimek B (ed) Grzimek's animal life encyclopedia. Van Nostrand Reinhold, New York, pp 236-254

Nader IA (1996) Distribution and status of five species of predators in Saudi Arabia. J Wildl Res 1(2):210-214

Nel JAJ (1978) Notes on the food and foraging behavior of the bat-eared fox, *Otocyon megalotis*. Bull Carnegie Mus Nat Hist 6:132-137

Nel JAJ, Bester MH (1983) Communication in the southern bat-eared fox *Octocyon m. megalotis* (Desmarest, 1822). Z Saugetierkd 48(5):277-290

Nel JAJ, Bothma JduP (1983) Scent marking and midden use by aardwolves (*Proteles cristatus*) in the Namib Desert. Afr J Ecol 21:25-39

Nel JAJ, Loutit R (1986) The diet of the black-backed jackals, *Canis mesomelas*, on the Namib Desert coast. Cimbebasia A 8(11):91-96

Nel JAJ, Mills MGL, Van Aarde RJ (1984) Fluctuating group size in bat-eared foxes (*Otocyon m. megalotis*) in the south-western Kalahari. J Zool (Lond) 203(2):294-298

Newsome AE, Corbett LK, Catling PC, Burt RJ (1983a) The feeding ecology of the dingo I. Stomach contents from trapping in south-east Australia, and the non-target wildlife also caught in dingo traps. Aust Wildl Res 10:477-486

Newsome AE, Catling PC, Corbett LK (1983b) The feeding ecology of the dingo. II. Dietary and numerical relationships with fluctuating prey populations in south-eastern Australia. Aust J Ecol 8:345-366

Norton PM, Lawson AB, Henley SR, Avery G (1986) Prey of leopards in four mountainous areas of the south-western Cape Province. S Afr J Wildl Res 16(2):47-52

Nowell K, Jackson P (1995) New red list categories for wild cats. Cat News 23:21-27

Nowell K, Jackson P (1996) Wild cats. IUCN Publication Services, Cambridge

O'Brien SJ (1994) The cheetahs' conservation controversy. Conserv Biol 8(4):1153-1155

Oli MK, Taylor IR, Rogers ME (1993) Diet of the snow leopard (*Panthera uncia*) in the Annapurna Conservation Area, Nepal. J Zool (Lond) 231:365-370

Owens DD, Owens MJ (1979a) Communal denning and clan associations in brown hyaenas (*Hyaena brunnea*, Thunberg) of the central Kalahari Desert. Afr J Ecol 17:35-44

Owens DD, Owens MJ (1979b) Notes on social organization and behavior in brown hyaenas (*Hyaena brunnea*). J Mammal 60:405-408

Owens DD, Owens MJ (1984) Helping behaviour in brown hyaenas. Nature 308:843-845

Owens MJ, Owens DD (1978) Feeding ecology and its influence on social organization in brown hyaenas (*Hyaena brunnea* Thunberg) of the central Kalahari Desert. East Afr Wildl J 16:113-135

Owens MJ, Owens DD (1984) Cry of the Kalahari. Houghton Mifflin, Boston

Packer C, Kock R (1995) Serengeti lions recovering from canine distemper epidemic. Cat News 23:9

Packer C, Pusey AE (1983) Adaptations of female lions to infanticide by incoming males. Am Nat 121(5):716-728

Packer C, Pusey AE (1987) Intrasexual cooperation and the sex ratio in African lions. Am Nat 130(4):636-642

Palgrave KC (1993) Trees of southern Africa, 2nd edn. Struik, Cape Town
Palmer R, Fairall N (1988) Caracal and African wild cat diet in the Karoo National Park and the implications therefore for hyrax. S Afr J Wildl Res 18(1):30-34
Paquet PC (1991) Scent-marking behavior of sympatric wolves (*Canis lupus*) and coyotes (*Canis latrans*) in Riding Mountain National Park. Can J Zool 69:1721-1727
Phillips JA (1993) Bone consumption by cheetahs at undisturbed kills : evidence for a lack of focal-palatine erosion. J Mammal 74(2):487-492
Pienaar UdeV (1960) Annual report of the Biologist, 1958/1959. Koedoe 3:1-205
Pienaar UdeV (1969) Predator-prey relationships amongst the larger mammals of the Kruger National Park. Koedoe 12:108-184
Poché RM, Evans SJ, Sultana P, Haque ME, Sterner R, Siddique MA (1987) Notes on the golden jackal (*Canis aureus*) in Bangladesh. Mammalia 51(2):259-270
Pournelle GH (1965) Observations on birth and early development of the spotted hyaena. J Mammal 46:503
Pringle JA, Pringle VL (1979) Observations on the lynx *Felis caracal* in the Bedford district. S Afr J Zool 14:1-4
Racey PA, Skinner JD (1979) Endocrine aspects of sexual mimicry in spotted hyaenas *Crocuta crocuta*. J Zool (Lond) 187:315-326
Rasa OAE (1973) Marking behaviour and its social significance in the African dwarf mongoose, *Helogale undulata rufula*. Z Tierpsychol 32:293-318
Rasa OAE (1986) Coordinated vigilance in dwarf mongoose family groups: the "Watchman's Song" hypothesis and the costs of guarding. Ethology 71:340-344
Rasa OAE (1987a) The dwarf mongoose : a study of behaviour and social structure in relation to ecology in a small, social carnivore. Adv Study Behav 17:121-163
Rasa OAE (1987b) Vigilance behaviour in dwarf mongooses: selfish or altruistic? S Afr J Sci. 83:587-590
Rasa OAE (1989a) Behavioural parameters of vigilance in the dwarf mongoose: social acquisition of a sex-based role. Behaviour 110 (1-4): 125-145
Rasa OAE (1989b) The costs and effectiveness of vigilance behaviour in the dwarf mongoose: implications for fitness and optimal group size. Ethol Ecol Evol 1:265-282
Rasa OAE (1994) Altruistic infant care or infanticide: the dwarf mongoose's dilemma. In: Parmagiani S, Vom Saal FS (eds) Infanticide and parental care. Academic Press, Harwood, pp 301-320
Rasa OAE, Wenhold BA, Howard P, Marais A, Pallett J (1992) Reproduction in the yellow mongoose revisited. S Afr J Zool 27(4): 192-195
Rautenbach IL, Nel JAJ (1978) Coexistence in Transvaal Carnivora. Bull Carnegie Mus Nat Hist 6:138-145
Richardson PRK (1985) The social behaviour and ecology of the aardwolf *Proteles cristatus* (Sparrman, 1783) in relation to its food resources. PhD Dissertation, University of Oxford, Oxford
Richardson PRK (1987a) Aardwolf: the most specialized myrmecophagous mammal? S Afr J Sci 83:643-646
Richardson PRK (1987b) Food consumption and seasonal variation in the diet of the aardwolf *Proteles cristatus* in southern Africa. Z Saugetierkd 52(5):307-325
Richardson PRK (1987c) Aardwolf mating system: overt cuckoldry in an apparently monogamous mammal. S Afr J Sci 83:405-410
Richardson PRK (1990) Scent marking and territoriality in the aardwolf. In: Macdonald DW, Müller-Schwarze D, Natynczuk SE (eds) Chemical signals in vertebrates. Oxford University Press, Oxford, pp 378-387
Richardson PRK (1991) Territorial significance of scent-marking during the non-mating season in the aardwolf *Proteles cristatus* (Carnivora: Protelidae). Ethology 87:9-27
Richardson PRK, Coetzee M (1988) Mate desertion in response to female promiscuity in the socially monogamous aardwolf *Proteles cristatus*. S Afr J Zool 23(4):306-308
Richardson PRK, Levitan CD (1994) Tolerance of aardwolves to defence secretions of *Trinervitermes trinervoides*. J Mammal 75(1):84-91

Rieger I (1979a) A review of the biology of striped hyaenas, *Hyaena hyaena* (Linné, 1758). Saugetierkd Mitt 27:81-95

Rieger I (1979b) Scent rubbing in carnivores. Carnivore 2(1):17-25

Rodgers WA (1974) The lion (*Panthera leo*, Linn.) population of the eastern Selous Game Reserve. East Afr Wildl J 12:313-317

Rood JP (1990) Group size, survival, reproduction, and routes to breeding in dwarf mongooses. Anim Behav 39: 566-572

Rudnai J (1973) Reproductive biology of lions (*Panthera leo massaica* Neumann) in Nairobi National Park. East Afr Wildl J 11:241-253

Saba ARK (1979) Predator-prey interaction: a case-study in the Masai-Mara Game Reserve, Kenya. In: Ajayi SS, Halstead LB (eds) Wildlife management in savannah woodland. Taylor and Francis, London, pp 41-49

Sandell M (1989) The mating tactics and spacing patterns of solitary carnivores. In: Gittleman JL (ed) Carnivore behavior, ecology and evolution. Chapman and Hall, London, pp 164-182

Schaller GB (1972) The Serengeti lion. University of Chicago Press, Chicago

Schaller GB, Lawther GR (1969) The relevance of carnivore behavior to the study of early hominids. Southwest J Anthropol 25(4):307-341

Scheel D, Packer C (1991) Group hunting behaviour of lions: a search for cooperation. Anim Behav 41:697-709

Scheel D, Packer C (1995) Variation in predation by lions: tracking a movable feast. In: Sinclair ARE, Arcese P (eds) Serengeti II: dynamics, management and conservation of an ecosystem. University of Chicago Press, Chicago, pp 299-314

Scheepers JL, Venzke KAE (1995) Attempts to reintroduce African wild dogs *Lycaon pictus* into Etosha National Park, Namibia. S Afr J Wildl Res 25(4): 138-140

Schmidt-Nielsen K (1964) Desert animals. Clarendon Press, Oxford

Scott J (1985) The leopard's tale. Elm Tree Books, London

Seely MK, Griffin M (1986) Animals of the Namib Desert: interactions with their physical environment. Rev Zool Afr 100:47-61

Seidensticker J (1977) Notes on early maternal behavior of the leopard. Mammalia 41(1):111-113

Senzota RBM (1990) Plains gerbils *Tatera robusta* as prey of golden jackals and owls in the Serengeti National Park. Acta Theriol 35(1-2):157-161

Siegfried WR (1984) An analysis of faecal pellets of the brown hyaena on the Namib coast. S Afr J Zool 19(1):61

Sillero-Zubiri C, Gottelli D (1995) Diet and feeding behavior of Ethiopian wolves (*Canis simensis*). J Mammal 76(2):531-541

Simpson CD (1964) Notes on the banded mongoose, *Mungos mungo* (Gmelin). Arnold Rhod 1(19):1-8

Sinclair ARE (1977) The African buffalo. University of Chicago Press, Chicago

Singh HS (1995) Asiatic lion census. Cat News 23:10

Skinner JD (1976) Ecology of the brown hyaena *Hyaena brunnea* in the Transvaal with a distribution map for southern Africa. S Afr J Sci 72:262-269

Skinner JD (1979) Feeding behaviour in caracal (*Felis caracal*). J Zool (Lond) 189:523-525

Skinner JD, Funston PJ, Van Aarde RJ, Van Dyk G, Haupt MA (1992) Diet of spotted hyaenas in some mesic and arid southern African game reserves adjoining farmland. S Afr J Wildl Res 22(4):119-121

Skinner JD, Henschel JR, Van Jaarsveld AS (1986) Bone-collecting habits of spotted hyaenas *Crocuta crocuta* in the Kruger National Park. S Afr J Zool 21:303-308

Skinner JD, Ilani G (1979) The striped hyaena *Hyaena hyaena* of the Judean and Negev Deserts and a comparison with the brown hyaena *Hyaena brunnea*. Isr J Zool 28:229-232

Skinner JD, Smithers RHN (1990) The mammals of the southern African subregion, 2nd edn. University of Pretoria, Pretoria

Skinner JD, Van Aarde RJ (1981) The distribution and ecology of the brown hyaena *Hyaena brunnea* and spotted hyaena *Crocuta crocuta* in the central Namib Desert. Madoqua 12(4):231-239

Skinner JD, Van Aarde RJ (1986) The use of space by the aardwolf *Proteles cristatus.* J Zool (Lond) 209:299-301

Skinner JD, Van Aarde RJ, Van Jaarsveld AS (1984) Adaptations in three species of large mammals (*Antidorcas marsupialis, Hystrix africaeaustralis, Hyaena brunnea*) to arid environments. S Afr J Zool 19:82-86

Smith RM (1978) Movement patterns and feeding behavior of the leopard in the Rhodes Matopos National Park, Rhodesia. Carnivore 1 (3-4):58-69

Smithers RHN (1971) The mammals of Botswana. Museum Memoirs, vol 4. National Museums, Salisbury

Smithers RHN (1983) The mammals of the southern African subregion. University of Pretoria, Pretoria

Smuts GL (1978a) More sex ratio data on lions. Carnivore 1(2):1

Smuts GL (1978b) Effects of population reduction on the travels and reproduction of lions in Kruger National Park. Carnivore 1(2):61-72

Smuts GL (1979) Diet of lions and spotted hyaenas assessed from stomach contents. S Afr J Wildl Res 9:19-25

Smuts GL (1982) Lion. Macmillan, Johannesburg

Smuts GL, Hanks J, Whyte IJ (1978) Reproduction and social organization of lions from the Kruger National Park. Carnivore 1(1):17-28

Stander PE (1991) Demography of lions in the Etosha National Park, Namibia. Madoqua 18(1):1-9

Stander PE (1992a) Cooperative hunting in lions: the role of the individual. Behav Ecol Sociobiol 29:445-454

Stander PE (1992b) Foraging dynamics of lions in a semi-arid environment. Can J Zool 70:8-21

Stephens DW, Krebs JR (1986) Foraging theory. Princeton University Press, Princeton

Stuart CT (1976) Diet of the black-backed jackal *Canis mesomelas* in the central Namib Desert, South West Africa. Zool Afr 11(1):193-205

Stuart CT, Shaughnessy PD (1984) Content of *Hyaena brunnea* and *Canis mesomelas* scats from southern coastal Namibia. Mammalia 48(4):611-612

Sunquist ME (1983) Dispersal of three radio-tagged leopards. J Mammal 64(2):337-341

Sunquist ME, Sunquist FC (1989) Ecological constraints on predation by large felids. In: Gittleman JL (ed) Carnivore behavior, ecology and evolution. Chapman and Hall, London, pp 283-301

Sutcliffe AJ (1970) Spotted hyaena: crusher, gnawer, digester and collector of bones. Nature 227:1110-1113

Tilson RL, Henschel JR (1986) Spatial arrangement of spotted hyaena groups in a desert environment, Namibia. Afr J Ecol 24:173-180

Tilson RL, Von Blottnitz F, Henschel JR (1980) Prey selection by spotted hyaenas (*Crocuta crocuta*) in the Namib Desert. Madoqua 12(1):41-49

Turnbull-Kemp P (1967) The leopard. Timmins, Cape Town

Turner A (1993) New fossil carnivore remains from Swartkrans. In : Brain CK (ed) Swartkrans, a cave's chronicle of early man. Transvaal Mus Monogr 8:1-270

Turner M, Watson M (1964) A census of game in the Ngorongoro Crater. East Afr Wildl J 2:165-168

Tyler S (1975) The Simien fox. Wildlife 17(12):564-565

Van Aarde RJ, Van Dyk A (1986) Inheritance of the king coat colour pattern in cheetahs *Acinonyx jubatus.* J Zool (Lond) [A] 209:573-578

Van Heerden J (1980) The transmission of *Babesia canis* to the wild dog *Lycaon pictus* (Temminck) and the black-backed jackal *Canis mesomelas* Schreber. J S Afr Vet Assoc 51(2):119-120

Van Heerden J (1981) The role of integumental glands in the social and mating behaviour of the hunting dog *Lycaon pictus* (Temminck, 1820). Onderstepoort J Vet Res 48:19-21

Van Heerden J, Mills MGL, Van Vuuren MJ, Kelly PJ, Dreyer MJ (1995) An investigation into the health status and diseases of wild dogs (*Lycaon pictus*) in the Kruger National Park. J S Afr Vet Assoc 66(1):18-27

Van Jaarsveld AS, Skinner JD (1987) Spotted hyaena monomorphism : an adaptive "phallusy"? S Afr J Sci 83:612-615

Van Jaarsveld AS, Henschel JR, Skinner JD (1987) Improved age estimation in spotted hyaenas (*Crocuta crocuta*). J Zool (Lond) 231:758-762

Van Jaarsveld AS, Skinner JD, Lindeque M (1988) Growth, development and parental investment in the spotted hyaena, *Crocuta crocuta*. J Zool (Lond) 216:45-53

Van Lawick H, Van Lawick-Goodall J (1970) Innocent killers. Collins, London

Van Valkenburgh B (1996) Feeding behavior in free-ranging African carnivores. J Mammal 77(1):240-254

Verberne G, De Boer J (1976) Chemocommunication among domestic cats, mediated by the olfactory and vomeronasal senses. Z Tierpsychol 42:86-109

Verberne G, Leyhausen P (1976) Marking behaviour of some Viverridae and Felidae: time-interval analysis of the marking pattern. Behaviour 58 (3-4):192-253

Viljoen S, Davis DHS (1973) Notes on stomach contents analyses of various carnivores in southern Africa (Mammalia: Carnivora). Ann Transvaal Mus 28(16):353-363

Von Gauthier-Pilters H (1966) Einige Beobachtungen über das Spielverhalten beim Fenek (*Fennecus zerda* Zimm.). Z Saugetierkd 31:337-350

Von Gauthier-Pilters H (1967) The fennec. Afr Wildl 21:117-125

Von Koenig L (1970) Zur Fortplanzung und Jugendentwicklung des Wüstenfuchses (*Fennecus zerda* Zimm.1780). Z Tierpsychol 27:205-246

Von Richter W (1972) Remarks on present distribution and abundance of some South African carnivores. J S Afr Wildl Manage Assoc 2:9-16

Wagner FH (1980) Wildlife of the deserts. Abrams, New York

Walker EP (1983) Walker's mammals of the world, vol 2, 4th edn. Johns Hopkins University Press, London

Wandry R (1975) Contribution to the study of the social behaviour of captive golden jackals (*Canis aureus* L.). Z Tierpsychol 39(1-5):365-402

Waser PM, Keane B, Creel SR, Elliott LF, Minchella DJ (1994) Possible male coalitions in a solitary mongoose. Anim Behav 47:289-294

Wells ME (1968) A comparison of the reproductive tracts of *Crocuta crocuta*, *Hyaena hyaena* and *Proteles cristatus*. East Afr Wildl J 6:63-70

Wenhold BA, Rasa OAE (1994) Territorial marking in the yellow mongoose *Cynictis penicillata:* sexual advertisement for subordinates ? Z Saugetierkd 59:129-138

Werdelin L, Solounias N (1991) The Hyaenidae: taxonomy, systematics and evolution. Fossils Strata 30:3-104

Whitehead H (1990) Rules for roving males. J Theor Biol 145:355-368

Wilson EO (1975) Sociobiology: the new synthesis. Belknap Press of Harvard University Press, Cambridge

Wozencraft WC (1993) Order Carnivora. In: Wilson DE, Reeder DM (eds) Mammal species of the world, 2nd edn. Smithsonian Institution Press, Washington, DC, pp 279-348

Wright BS (1960) Predation on big game in East Africa. J Wildl Manage 24(1):1-15

Wyman J (1967) The jackals of the Serengeti. Animals (Lond) 10:79-83

Zumpt IF (1968) The feeding habits of the yellow mongoose, *Cynictis penicillata*, the suricate, *Suricata suricatta* and the Cape ground squirrel, *Xerus inauris*. J S Afr Vet Med Assoc 39:89-91

Genus and Species Index

Subject Index

Printing: Mercedesdruck, Berlin
Binding: Buchbinderei Lüderitz & Bauer, Berlin